Waves in the Ocean and Atmosphere

T0202794

Springer

Berlin
Heidelberg
New York
Hong Kong
London
Milan
Paris
Tokyo

Joseph Pedlosky

Waves
in the Ocean
and Atmosphere

Introduction to Wave Dynamics

With 95 Figures

 Springer

Author

Dr. Joseph Pedlosky
Woods Hole Oceanographic Institution
Department of Physical Oceanography
Clark 363 MS 21
Woods Hole, MA 02543
USA

e-mail: *jpedlosky@whoi.edu*

ISBN 978-3-642-05564-5

Library of Congress Cataloging-in-Publication Data Applied For

Bibliographic information published by Die Deutsche Bibliothek
Die Deutsche Bibliothek lists this publication in the Deutsche Nationalbibliografie;
detailed bibliographic data is available in the Internet at http://dnb.ddb.de

Springer-Verlag Berlin Heidelberg New York
a member of BertelsmannSpringer Science+Business Media GmbH
http://www.springer.de
© Springer-Verlag Berlin Heidelberg 2010
Printed in Germany

Cover Design: Erich Kirchner, Heidelberg
Dataconversion: Büro Stasch, Bayreuth

Printed on acid-free paper – 32/3140 – 5 4 3 2 1 0

Preface

For over twenty years, the Joint Program in Physical Oceanography of MIT and the Woods Hole Oceanographic Institution has based its education program on a series of core courses in Geophysical Fluid Dynamics and Physical Oceanography. One of the central courses in the Core is one on wave theory, tailored to meet the needs of both physical oceanography and meteorology students. I have had the pleasure of teaching the course for a number of years, and I have particularly enjoyed the response of the students to their exposure to the fascination of wave phenomena and theory.

This book is a reworking of course notes that I have prepared for the students, and I was encouraged by their enthusiastic response to the notes to reach a larger audience with this material. The emphasis, both in the course and in this text, is twofold: the development of the basic ideas of wave theory and the description of specific types of waves of special interest to oceanographers and meteorologists. Throughout the course, each wave type is introduced both for its own intrinsic interest and importance and as a vehicle for illustrating some general concept in the theory of waves. Topics covered range from small-scale surface gravity waves to large-scale planetary vorticity waves. Concepts such as energy transmission, reflection, potential vorticity, the equatorial wave guide, and normal modes are introduced one step at a time in the context of specific physical phenomena. Many topics associated with steady flows are also illustrated to great benefit through a consideration of wave theory and topics such as geostrophic adjustment, the transformation of scale under reflection, and wave-mean flow interaction. These are natural links between the material of this course and theories of steady currents in the atmosphere and oceans.

The subject of wave dynamics is an old one, and so much of the material in this book can be found in texts, some of them classical, and well-known papers on certain aspects of the subject. It would be hard to claim originality for the standard ideas and concepts, some of which, like tidal theory, can be traced back to the nineteenth century. Other more recent ideas, such as the asymptotic approach to slowly varying wave theory found in texts such as Whitham's or Lighthill's, have been borrowed and employed to illuminate the subject. In each case, references at the end of the text for each section indicate the sources that I found particularly useful. What I have tried to do in the course and in this text is to weave those ideas together in a way that I personally believe makes the subject as accessible as possible to first-year graduate students. Indeed, I have tried to retain some of the informality in the text of the original notes. The text is composed of twenty one "lectures," and the reader will note from time to time certain questions posed didactically to the student and certain challenges to the reader to obtain some results independently. A series of problem sets, which the students found helpful, are placed at the end of the text.

My teaching and research at the Woods Hole Oceanographic Institution has been generously supported by the Henry L. and Grace Doherty chair in Physical Oceanography for which I am delighted to express my appreciation. I also am happy to express my gratitude for years of support from the National Science Foundation, which recognizes the inextricably linked character of research and teaching.

The waves course has been fun to teach. The fascination of the material seems to naturally engage the curiosity of the students and it is to them, collectively, that this book is dedicated.

Joseph Pedlosky

Woods Hole
May 05, 2003

Contents

Introduction

A course on wave motions for oceanographers and meteorologists has (at least) two purposes.

The first is to discuss the important types of waves that occur in the atmosphere and oceans, in order to understand their *properties, behavior,* and how to *include them in our overall picture of the ocean and atmosphere.* There are a large number of such waves, each with different physics, and it will be impossible to discuss all of them exhaustively.

At the same time, a second purpose of the course is to develop the theory and concepts of waves themselves. What are waves? What does it mean for a wave to move? What does the wave do to the medium in which it propagates, and vice-versa? How do waves (if they do) interact with one another? How do they arise? All of these are good and fundamental questions.

In order to deal with both of these goals, the course will describe a series of different waves and use each wave type to describe a different aspect of basic wave theory. It will then be up to you to form the necessary connections and generalize the ideas to all waves, at least on a heuristic basis. This will require you to sometimes retroactively apply some new ideas developed in the discussion of wave type B, for example, back to the application of wave type A discussed previously in the course.

In general, the physical ingredients will be stratification and rotation. But first, what is a wave?

There is no definition of a wave that is simple and general enough to be useful, but in a rough way we can think of a wave as:

> *A moving signal, typically moving at a rate distinct from the motion of the medium.*

A good example is the "wave" in a sports stadium. The pattern of the wave moves rapidly around the park. The signal consists in the *cooperative* motion of individuals. The signal moves a much greater distance than the motion of any individual. In fact, while each person moves only up and down, the signal moves laterally (until it gets to the costly box seats where it frequently dissipates).

Similarly in a fluid whose signal could be an acoustic pressure pulse, the surface elevation of the ocean in a gravity wave, the rippling of the 500 mb surface in the troposphere due to a cyclone wave, or the distortion of the deep isopycnals in the thermocline due to internal gravity waves, the wave moves faster and further than the individual fluid elements. Thus, usually if

- u = the characteristic velocity of the fluid element in the wave, and
- c = the signal speed of the wave,

$$\frac{u}{c} \ll 1 \tag{1.1}$$

We shall see that this is also equivalent to the condition for the linearization of the mathematical description of the wave physics.

Wave Kinematics

Before discussing wave physics, it is useful to establish some basic ideas and notational definitions about the kinematics of waves. A more complete discussion can be found in the excellent texts by Lighthill (1975) and Whitham (1974).

For simple systems and for small amplitude waves (i.e., when we linearize) we often can find solutions to the equations of motion in the form of a *plane wave*. This usually requires the medium to be, at least locally on the scale of the wave, homogeneous. If $\phi(x_i,t)$ is a field variable such as pressure,

$$\phi(\vec{x},t) = \phi(x_i,t) = \mathrm{Re}\,A e^{i(\vec{K}\cdot\vec{x}-\omega t)} \tag{1.2}$$

where
- A = the wave amplitude (complex so it includes a constant phase factor),
- $\vec{K}$ = the wave vector,
- ω = the wave frequency, and
- Re implies that the real part of the following expression is taken.

We can define the variable phase of the wave θ as

$$\theta(\vec{x},t) = \vec{K}\cdot\vec{x} - \omega t = k_i x_i - \omega t \tag{1.3}$$

where the summation convention is implied in the second form, that is,

$$k_i x_i \Leftrightarrow \sum_{j=1}^{\max\dim} k_j x_j \tag{1.4}$$

In the simplest case, A, ω and k_j are constants.

This begs the question of why we should ever observe a disturbance with a single $K = \vec{K}$ and ω. To understand that we must do more work later on. But standing on a beach and looking at the swell approaching it appears often to be the first order description of the wave field and a naturally simple case.

Of course, by Fourier's theorem (look it up now) we can represent *any* shape by a superposition of such plane waves.

The function ϕ we have considered above is constant on the surfaces (planes, hence the name) on which θ is constant, i.e.,

$$k_i x_i - \omega t = \text{constant} \tag{1.5}$$

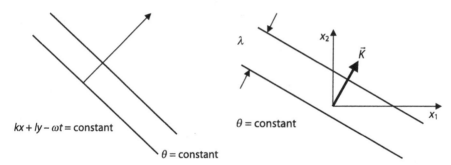

Fig. 1.1. Schematic of wave crest

Fig. 1.2. The plane wave showing the crests and wave vector and wavelength

In two dimensions, for example, these will be the lines

$$\vec{K} \cdot \vec{x} - \omega t = k_1 x_1 + k_2 x_2 - \omega t \equiv kx + ly - \omega t \tag{1.6}$$

(We will use the notation regularly, $k_1 = k$, $k_2 = l$, $k_3 = m$, in Cartesian coordinates). The directions of the lines of constant phase are given by the *normal* to those lines of constant θ (Fig. 1.1, Fig. 1.2), i.e.,

$$\nabla \theta = \nabla\left[\vec{K} \cdot \vec{x} - \omega t\right] = \vec{K} \tag{1.7}$$

or equivalently

$$\frac{\partial \theta}{\partial x_i} = \frac{\partial}{\partial x_i}\left(k_j x_j - \omega t\right) = k_j \frac{\partial x_j}{\partial x_i} = k_j \delta_{ij} = k_i \tag{1.8}$$

Define

$$K = \left|\vec{K}\right| \tag{1.9}$$

i.e., the magnitude of the wave vector. Then

$$\vec{K} \cdot \vec{x} = Ks \tag{1.10}$$

where s is the scalar distance perpendicular to the line of constant phase, for example the crests where ϕ is a maximum.

The plane wave is a spatially periodic function so that $\phi(Ks) = \phi(K[s + \lambda])$ where $K\lambda = 2\pi$, since

$$e^{i(Ks)} = e^{i(Ks+2\pi)}, \quad e^{i(2\pi)} \equiv 1$$

Thus,

$$\lambda = \frac{2\pi}{K} \tag{1.11}$$

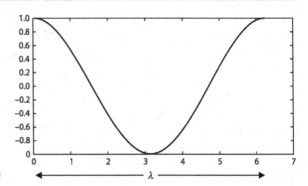

Fig. 1.3.
The wavelength of a plane wave

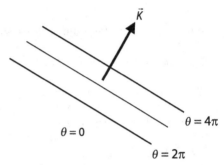

Fig. 1.4.
The increase of phase in the direction of the
wave vector

is the wavelength. It is the distance along the wave vector between two points of the
same phase (Fig. 1.3).

At any *fixed position*, the rate of change of the phase with time is given by

$$\frac{\partial \theta}{\partial t} = -\omega \tag{1.12}$$

ω is therefore the rate of *decrease* of phase (note: as crests arrive, moving parallel to
the wave vector K, the phase will decrease at a fixed point (see Fig. 1.4).

How long do we have to wait until the same phase appears? The shortest wait occurs
when a time T has passed such that $\omega T = 2\pi$. The time T is called the *wave period*,
and

$$T = \frac{2\pi}{\omega} \tag{1.13}$$

What is the speed of movement of the line of constant phase

$$\theta = k_j x_j - \omega t = Ks - \omega t \quad ? \tag{1.14}$$

Note that as t increases, s must increase to keep the phase constant (Fig. 1.5, 1.6).

$$\left. \frac{\partial s}{\partial t} \right)_{\theta} = -\frac{\partial \theta / \partial t}{\partial \theta / \partial s} = \frac{\omega}{K} \tag{1.15}$$

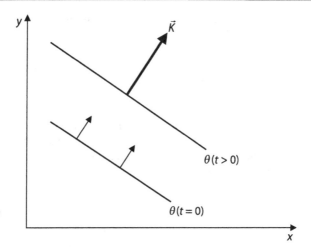

Fig. 1.5.
The movement with time of the line of constant phase in the direction of the wave vector

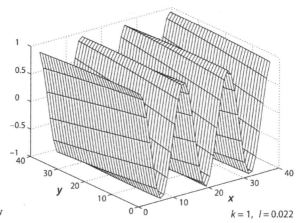

Fig. 1.6.
A plane wave in perspective view

$k = 1, l = 0.022$

Be sure you understand the reason for the appearance of the minus sign:

{At constant θ, $d\theta = 0 = Kds - \omega dt$, so that $ds/dt = \omega/K$}

We define the phase *speed* to be the speed of propagation of phase *in the direction of the wave vector.*

phase speed: $c = \omega/K$

Note that phase speed is not a vector. For example, in two dimensions the phase speed in the x-direction would be defined such that at fixed y,

$$d\theta = 0 = kdx - \omega dt \quad \text{or} \tag{1.16a}$$

$$c_x = \frac{\omega}{k} = -\frac{\partial\theta/\partial t}{\partial\theta/\partial x} \tag{1.16b}$$

Note that if the phase speed *were* a vector directed in the direction of **K**, its *x*-component would be

$$\vec{c} \cdot \hat{i} = \frac{\omega}{K} \frac{\vec{K}}{K} \cdot \hat{i} = \frac{\omega}{K^2} k \neq c_x \qquad (1.17)$$

Therefore, it is clear that the phase speed does not act like a vector, and this is a clue that this speed, by which the *pattern* of the wave propagates, may have less physical meaning that we would intuitively want to give to it.

Note that c_x is the speed with which the intersection of the moving phase line with the *x*-axis moves along the *x*-axis (Fig. 1.7):

$$c_x = \frac{c}{\cos \alpha}$$

and as α goes to $\pi/2$, c_x becomes infinitely large! This makes us suspicious that the phase may not be the messenger of physical entities like momentum and energy.

In an interval length *s* perpendicular to the surface of constant phase, the increase in phase divided by 2π gives us the number of crests in the interval. Thus, Fig. 1.8.

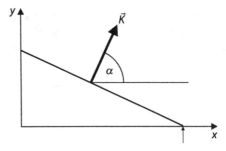

Fig. 1.7.
The *small arrow* shows the intersection point of the line of constant phase and the *x*-axis

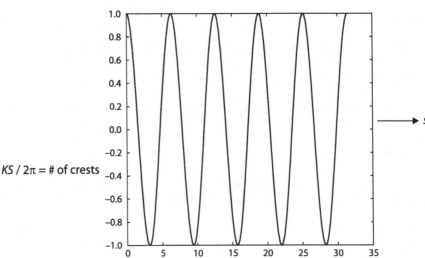

$KS / 2\pi = $ # of crests

Fig. 1.8. A plane wave and the number of crests along the coordinate *s*

Thus also, the increase in phase along the wave vector is

$$\Delta\theta = \int \frac{\partial\theta}{\partial s} ds = \int K ds \tag{1.18}$$

The more fundamental definitions have already been given; namely,

$$\vec{K} = \nabla\theta \tag{1.19a}$$

$$\omega = -\frac{\partial\theta}{\partial t} \tag{1.19b}$$

The former gives the spatial increase of phase, while the latter gives the temporal (decrease) of phase.

In all physical wave problems, the dynamics will impose, as we shall see, a relation between the wave vector and the frequency. This relation is called the *dispersion relation* (for reasons that will be made more clear later). The form of the dispersion relation can be written as:

$$\omega = \Omega(k_j) \tag{1.20}$$

Note that each wave vector has its own frequency. Often the frequency depends only on the magnitude of the wave vector, K, rather than its orientation, but this is not always the case. Up to now, the wave vector, the frequency, the phase speed and the dispersion relation have all been considered constants, i.e., *independent of space and time*.

Kinematic Generalization

Suppose the medium is not homogeneous. For example, gravity waves impinging on a beach see of varying depth as the waves run up the beach, acoustic waves see fluid of varying pressure and temperature as they propagate vertically, etc. Then a pure plane wave in which all attributes of the wave are constant in space (and time) will not be a proper description of the wave field. Nevertheless, if the changes in the background occur on scales that are *long* and *slow* compared to the wavelength and period of the wave, a plane wave representation may be *locally* appropriate (Fig. 2.1). Even in a homogeneous medium, the wave might change its length if the wave is a superposition of plane waves (as we shall see later).

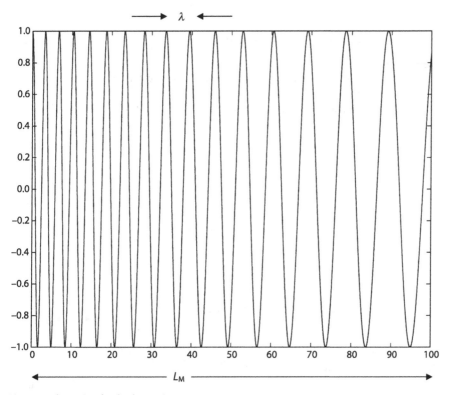

Fig. 2.1. Schematic of a slowly varying wave

Thus, locally the wave can still look like a plane wave if $\lambda / L_M \ll 1$. In that case, we might expect the wave to be described by the *form*:

$$\phi(\vec{x},t) = A(x,t)e^{i\theta(x,t)} \quad \text{(the real part of the expression is taken for granted),} \quad (2.1)$$

where A varies on the scale L_M while the phase varies on the scale λ. Thus,

$$\frac{1}{A}\frac{\partial A}{\partial x_i} = O\left(\frac{1}{L_M}\right) \tag{2.2a}$$

$$\frac{\partial \theta}{\partial x_i} = O\left(\frac{1}{\lambda}\right) \tag{2.2b}$$

so that

$$\nabla \phi = Ae^{i\theta}\nabla\theta + O\left(\frac{\lambda}{L_M}\right) \tag{2.3}$$

We *define* (guided by our experience with the plane wave):

$$\vec{K} = \nabla\theta \quad \text{local spatial increase of phase} \tag{2.4a}$$

$$-\omega = \frac{\partial\theta}{\partial t} \quad \text{local increase of phase with time} \tag{2.4b}$$

Since the wave vector is defined as the gradient of the scalar phase, it follows automatically that $\nabla \times \vec{K} = 0$.

Consider the increase of phase on the curve C_1 from point A to point B in Fig. 2.2:

$$n_{C_1} = \frac{1}{2\pi}\int_A^B \vec{K}\cdot d\vec{x} = \frac{1}{2\pi}\int_{C_1} \vec{K}\cdot d\vec{x} \tag{2.5}$$

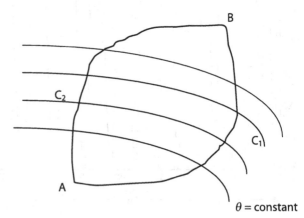

Fig. 2.2.
Counting crests on two paths
AC_1B and AC_2B

Now consider the same increase calculated on curve C_2:

$$n_{C_2} = \frac{1}{2\pi} \int_A^B \vec{K} \cdot d\vec{x} = \frac{1}{2\pi} \int_{C_2} \vec{K} \cdot d\vec{x} \tag{2.6}$$

The difference between them is

$$\begin{aligned} n_{C_1} - n_{C_2} &= \frac{1}{2\pi} \int_{C_1} - \int_{C_2} \vec{K} \cdot d\vec{x} = \frac{1}{2\pi} \oint_{C_{total}} \vec{K} \cdot d\vec{x} \\ &= \iint_A \nabla \times \vec{K} \cdot \hat{n} dA \\ &= 0 \end{aligned} \tag{2.7}$$

Here we have used Stokes theorem relating the line integral of the tangent component of K with the area integral of its curl over the area bounded by the closed contour composed of the sum of the two curves C_1 and C_2. Since the curl is zero, the two calculations for the increase of phase must be *independent of the curve used to do the calculation.*

Note that since

$$\vec{K} = \nabla \theta \tag{2.8a}$$

$$\omega = -\frac{\partial \theta}{\partial t} \tag{2.8b}$$

it follows *by definition* that

$$\frac{\partial \vec{K}}{\partial t} + \nabla \omega = 0 \tag{2.9}$$

in those cases where the wave vector and the wave frequency are slowly varying functions of space and time (i.e., where it is sensible to define wavelength and frequency).

To better understand the consequences of the above equation, consider the fixed line element AB in Fig. 2.3.

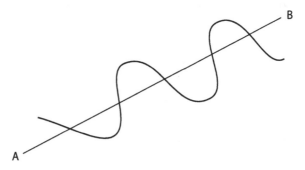

Fig. 2.3.
Conservation of crests along the line AB

Integrate the above conservation equation along the line element from A to B:

$$\frac{\partial}{\partial t}\int_A^B \vec{K}\cdot d\vec{x} + \int_A^B \nabla\omega\cdot d\vec{x} = 0 \tag{2.10}$$

Using our previous definitions, in particular that $Ks/2\pi$ is the number of crests in the interval s, it follows from the above that

$$\frac{\partial n_{AB}}{\partial t} = \frac{\omega(A)}{2\pi} - \frac{\omega(B)}{2\pi} \tag{2.11}$$

That is to say, the rate of change of the number of crests in the interval (A,B) is equal to the rate of inflow of crests at point A minus the outflow of crests at point B, since the frequency (divided by 2π) is equal to the number of crests crossing a point at each moment. E.g.,

$\omega(A)$ = rate of decrease of phase at point A (see Fig. 2.4)

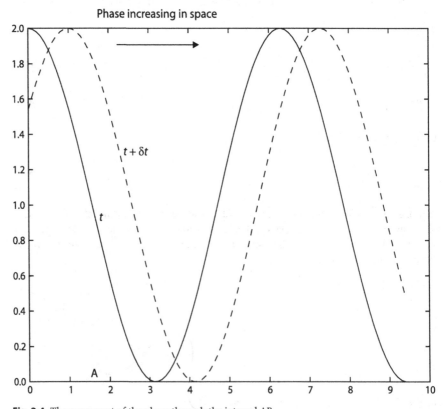

Fig. 2.4. The movement of the phase through the interval AB

We may think of this as a statement of the *conservation of wave crests*. Namely, the number of wave crests in a smoothly varying function ϕ as given above does not change. The number in any local region increases or decreases solely due to the arrival of preexisting crests, not to the creation or destruction of existing crests.

Now, let's suppose that we still have a local dispersion relation between frequency and wave number but that the relationship slowly changes on scales that are long compared with a wavelength or period due to changes, perhaps, in the nature of the medium in which the wave is embedded.

In that case, the natural generalization of the dispersion relation is

$$\omega = \Omega(k_j, x_i, t) \tag{2.12}$$

where the wave vector components and the frequency may themselves be functions of space and time (slowly), and the dispersion relation is *explicitly* dependent on space and time.

Thus,

$$\frac{\partial \omega}{\partial t} = \frac{\partial \Omega}{\partial t}\bigg)_{\vec{K},\vec{x}} + \frac{\partial \Omega}{\partial k_j}\frac{\partial k_j}{\partial t} \tag{2.13}$$

where the first term on the right-hand side is due to the *explicit* dependence of the dispersion relation on time, as might happen if the temperature of a region through which an acoustic wave were traveling were increasing with time.

We *define* the *group velocity* by the formula for each of its Cartesian components:

$$c_{g_j} = \frac{\partial \Omega}{\partial k_j} \tag{2.14}$$

for the component of the group velocity in the jth direction, or

$$\vec{c}_g = \nabla_{\vec{K}}\Omega \tag{2.15}$$

It follows from a fundamental theorem in vector analysis that since the phase is a scalar and the gradient operator is a vector, the group velocity is a true vector (distinct from the phase speed). That is, it follows the law of vector decomposition.

Since, by our earlier definitions

$$\frac{\partial k_j}{\partial t} = -\frac{\partial \omega}{\partial x_j} \tag{2.16}$$

we thus obtain

$$\frac{\partial \omega}{\partial t} = \frac{\partial \Omega}{\partial t} - \frac{\partial \Omega}{\partial k_j}\frac{\partial \omega}{\partial x_j} \tag{2.17}$$

It therefore follows that

$$\frac{\partial \omega}{\partial t} + \vec{c}_g \cdot \nabla \omega = \frac{\partial \Omega}{\partial t} \Leftarrow \text{explicit derivative with time} \tag{2.18}$$

Again, by similarly using

$$\frac{\partial k_i}{\partial t} + \frac{\partial \omega}{\partial x_i} = 0$$

it follows that

$$\frac{\partial k_i}{\partial t} + \frac{\partial \Omega}{\partial x_i} + \frac{\partial \Omega}{\partial k_j} \frac{\partial k_i}{\partial x_j} = 0 \quad \text{or} \tag{2.19a}$$

$$\frac{\partial k_i}{\partial t} + \frac{\partial \Omega}{\partial k_j} \frac{\partial k_i}{\partial x_j} = -\frac{\partial \Omega}{\partial x_i} \tag{2.19b}$$

Since the wave vector has no curl, it follows that

$$\frac{\partial k_i}{\partial x_j} = \frac{\partial k_j}{\partial x_i}$$

so the above equation can be rewritten:

$$\frac{\partial \vec{K}}{\partial t} + (\vec{c}_g \cdot \nabla)\vec{K} = -\nabla \Omega \Leftarrow \text{explicit dependence on space} \tag{2.20}$$

Note that the sum of derivatives on the left in the equations for the rate of change of wave vector and frequency are the rate of change for *an observer moving with the group velocity*.

So,

1. If the medium is independent of time, $\longrightarrow$ ω propagates with the group velocity;
2. If the medium is independent of space, $\longrightarrow$ K propagates with the group velocity.

If both (*1*) and (*2*) are true, both frequency and wave number propagate with the group velocity:

$$c_{g_i} = \frac{\partial \Omega}{\partial k_i}$$

This is a vector, and we see here that real wave attributes propagate with this velocity. If the dispersion relation is a function of space and/or time, the above equations tell us *how* the frequency and wave number change as we move with the group velocity following a wave. Further discussion can be found in Bretherton (1971) and Pedlosky (1987).

Example

We will soon see that free surface gravity waves (short enough so that rotation is unimportant but long enough so that the wavelength is large) compared to the depth have a dispersion relation:

$$\omega = k\sqrt{gH}$$

where H is the depth of the fluid and k is the wave number for this one-dimensional example (Fig. 2.5).

The phase speed and group velocity are equal in this case:

$$c_g = c = (gH)^{1/2}$$

If the depth is a function of x, then following a signal, since the dispersion relation is *independent of time,* the frequency will be constant for an observer moving with the velocity $c_g = c = (gH)^{1/2}$. For such an observer, with frequency constant, $k = \text{const.} / H^{1/2}$, which implies that the wave will grow shorter (larger k) as the wave enters shallow water. (It may become so short that it might break). Note that the observer, following a particular frequency moving with the group speed will proceed at a rate:

$$\frac{dx}{dt} = (gH(x))^{1/2} \tag{2.22}$$

For example, if $H(x)$ is of the form $H = H_0(1 - x / x_0)$ where x is measured positive shoreward from some offshore position a distance x_0 from the waterline (see Fig. 2.5), the signal corresponding to a given frequency will proceed onshore such that at a point x after an elapsed time t, the relationship between the elapsed time and its onshore progress is

$$t = 2x_0\left(1 - \sqrt{1 - x/x_0}\right)/(gH_0)^{1/2} \tag{2.23}$$

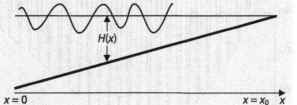

Fig. 2.5.
Water wave running up a sloped beach

The above kinematic discussion doesn't tell us how the amplitude of the wave propagates or, equivalently, how the energy in the wave moves. In some simple cases that are general enough to be of interest, we can actually describe how the amplitude and hence energy moves.

Consider the case of a *homogeneous* medium in which the governing equation for the wave function ϕ is of the form

$$\Pi(\partial/\partial t, \partial/\partial x_i)\phi(x_i,t)=0 \qquad (2.24)$$

where Π is a polynomial in the partial derivatives with respect to space and time. A simple example would be the Rossby wave equation:

$$\left(\frac{\partial^3}{\partial x^2 \partial t}+\frac{\partial^3}{\partial y^2 \partial t}+\beta\frac{\partial}{\partial x}\right)\phi=0 \qquad (2.25)$$

so that in this case,

$$\Pi=\frac{\partial}{\partial t}\left[\frac{\partial}{\partial x}\left(\frac{\partial}{\partial x}\right)+\frac{\partial}{\partial y}\left(\frac{\partial}{\partial y}\right)\right]+\beta\frac{\partial}{\partial x}$$

i.e., the polynomial in the partial derivatives are in respect of x, y and t.

Suppose we look for an *approximate* solution of the form

$$\phi = Ae^{i\theta}$$

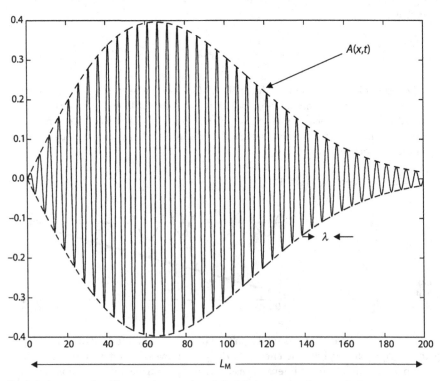

Fig. 2.6. A wave packet. The wave has wavelength λ while its *envelope* has a scale L_M

where A, k and ω are slowly varying functions of time, i.e., where the solution has the form of a one-dimensional *wave packet* (see Fig. 2.6), then

$$\frac{\partial \phi}{\partial x} = \left(i \frac{\partial \theta}{\partial x} A + \frac{\partial A}{\partial x} \right) e^{i\theta}$$
$$= \left(ikA + \frac{\partial A}{\partial x} \right) e^{i\theta} \quad, \text{etc.}$$

(2.26)

or

$$\Pi \phi = 0 \Rightarrow \Pi \left(-i\omega + \frac{\partial}{\partial t}, ik + \frac{\partial}{\partial x} \right) A = 0$$

(2.27)

Expanding the polynomial using the fact that the time and space derivatives of A are small compared to ω and k,

$$\Pi(-i\omega, ik) + \frac{\partial \Pi}{\partial(-i\omega)} \frac{\partial A}{\partial t} + \frac{\partial \Pi}{\partial(ik)} \frac{\partial A}{\partial x} = 0$$

(2.28)

The dispersion relation for plane waves comes from the disappearance of the first term (which is the dominant one), namely

$$\Pi(-i\omega, ik) = 0 \longrightarrow Linear\ dispersion\ relation$$

(2.29)

In the case above, this yields $\omega = -\beta / k$.
When this dispersion relation is satisfied, the remaining term yields the condition:

$$\frac{\partial A}{\partial t} - \frac{\partial \Pi / \partial k}{\partial \Pi / \partial \omega} \frac{\partial A}{\partial x} = 0$$

(2.30)

where the derivatives of Π in the equation occur when Π is evaluated as a function of frequency and wave number as in Eq. 2.29.
Since

$$\frac{\partial \Pi / \partial k}{\partial \Pi / \partial \omega} = -\frac{\partial \omega}{\partial k} \bigg)_{\Omega}$$

(2.31)

it follows that

$$\frac{\partial A}{\partial t} + c_g \frac{\partial A}{\partial x} = 0$$

(2.32)

Thus, the amplitude (and we can suppose) energy will propagate with the group velocity and not the phase speed. Where the *envelope (that is A)* of the wave goes, that is where the energy is. There is clearly no energy outside the wave envelope.

The reader should calculate the group velocity for this simple case of one-dimensional Rossby waves to see that the group and phase velocities are not the same. Similarly, the argument presented here can be extended to any number of dimensions (try it).

It is also clear that one might be able to use similar ideas for inhomogeneous media.

Once again we see here the physical primacy of the group velocity over the phase speed for the propagation of physical attributes of the wave.

Equations of Motion; Surface Gravity Waves

For a rotating stratified fluid, the general equations of motion can be written as:

1. Momentum equation:

$$\rho\left[\frac{d\vec{u}}{dt}+2\Omega\times\vec{u}\right]=-\nabla p+\mu\nabla^2\vec{u}+\kappa\nabla(\nabla\cdot\vec{u}) \quad \text{(if } \mu \text{ constant, } \kappa \text{ is second viscosity)} \quad (3.1)$$

2. Mass conservation:

$$\frac{\partial\rho}{\partial t}+\nabla\cdot(\rho\vec{u})=0 \quad ; \text{and} \tag{3.2}$$

3. Thermodynamic energy equation:

$$\frac{ds}{dt}=H$$

where s is specific entropy and H is the nonreversible heat addition. This can be re-written, assuming that s is a thermodynamic function of p and ρ,

$$c_p\frac{dT}{dt}-\frac{\alpha T}{\rho}\frac{dp}{dt}=\Phi+\frac{k}{\rho}\nabla^2 T+Q+\kappa(\nabla\cdot\vec{u})^2\equiv H \tag{3.3}$$

Here, T is temperature, c_p is the specific heat at constant pressure, α is the coefficient of thermal expansion, and Φ is the dissipation function, i.e., the frictional transformation of mechanical to thermal energy. If τ_{ij} is the stress tensor and e_{ij} is the rate of the strain tensor, $\Phi=\tau_{ij}e_{ij}$ (sums implied). Note that

$$\alpha=-\frac{1}{\rho}\left(\frac{\partial\rho}{\partial T}\right)_p \tag{3.4}$$

For a perfect gas with a state equation $p=\rho RT$, the thermodynamic equation is usually written in terms of the *potential temperature*:

$$\Theta=T\left(\frac{p_o}{p}\right)^{R/c_p}$$

so that the thermodynamic equation becomes

$$\frac{1}{\Theta}\frac{d\Theta}{dt} = \frac{H}{c_p T} \tag{3.5}$$

while for an incompressible liquid we can approximate the thermodynamic equation with

$$\frac{d\rho}{dt} = -\frac{\alpha\rho}{c_p}H \tag{3.6}$$

Here we have used the approximate state equation $\rho = \rho_0(1 - \alpha(T - T_0))$ to relate temperature in the thermodynamic equation to density. Be sure to note that when we make the approximation of incompressibility in the mass equation ($\nabla \cdot \vec{u} \approx 0$), this does *not* imply that $d\rho/dt = 0$ is the governing equation for density. *Only if* the dissipation H can be neglected will that be true. That is a separate physical statement about the adiabatic nature of the motion quite apart from the issue of compressibility. For a compressible fluid, we would have, instead of $d\rho/dt = 0$, the statement $ds/dt = 0$. For a detailed discussion of the formulation of these equations, especially the thermodynamics, see Batchelor (1967).

First Wave Example: Surface Gravity Waves

Perhaps the most familiar of waves in the ocean are the waves we see on the surface, either from a ship or from the beach (or from the air). These are waves on the interface between the water and the air (Fig. 3.1). The latter is so light compared with the former that we will approximate the air as having zero density to eliminate any dynamical interaction with the air to begin with. Theories of wave *generation* must include that coupling.

Consider a layer of liquid of uniform density and uniform depth. We suppose the scale of the motion is small enough to be able to ignore the Earth's rotation and the motion is small enough to be able to linearize this motion. In all such cases, we need to ask ourselves whether these statements are sensible, and if so, for what range of parameters? That is, if we ignore rotation, is there a limit, for example on the size of the wave for which that is appropriate? We might already know, for example, that the tides, which are a gravity wave response to the sun and the moon, do feel the effects of the Earth's rotation, but, of course, they are of planetary scale.

1. Can we ignore rotation, friction and nonlinearity?
   - To ignore rotation, compare $\partial/\partial t$ with $\Omega \longrightarrow$ this implies that we need $\omega \gg \Omega$.
   - To ignore friction, compare $\partial/\partial t$ with μk^2, where k is a typical value of wavenumber $\longrightarrow \omega \gg \mu k^2$.

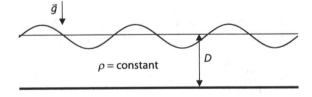

Fig. 3.1.
The homogeneous layer of
fluid supporting surface
gravity waves

- To ignore nonlinearity, compare $\partial/\partial t$ with respect to $\bar{u}\cdot\nabla \longrightarrow \omega \gg uk$ or $c \gg u \Longleftarrow$: this is the condition that the disturbance be wave-like, i.e., that the signal is carried by the wave rather than the advective motion of the fluid.

2. Can we treat the fluid as incompressible?

Assume we can linearize. Suppose the motion is adiabatic. In general then, we have

$$\frac{ds}{dt}=0, \quad s=s(p,\rho) \tag{3.7}$$

with the linearization

$$\frac{\partial s}{\partial t}=0=\frac{\partial s}{\partial p}\frac{\partial p}{\partial t}+\frac{\partial s}{\partial \rho}\frac{\partial \rho}{\partial t} \tag{3.8}$$

Thus,

$$\frac{\partial p}{\partial t}=-\frac{\partial s/\partial \rho}{\partial s/\partial p}\frac{\partial \rho}{\partial t}=\left(\frac{\partial p}{\partial \rho}\right)_s\frac{\partial \rho}{\partial t} \tag{3.9}$$

From the theory of acoustics we know (or we can easily find out) that the speed of sound in any medium is in fact given by the adiabatic compressibility of the medium. That implies that if c_a is the speed of sound in a fluid,

$$c_a^2=\left(\frac{\partial p}{\partial \rho}\right)_s$$

(One of the few scientific mistakes Newton made was to imagine that the speed of sound was this derivative at constant temperature and not entropy).

So we have the estimate for the relation between a perturbation in the density and the perturbation of the pressure:

$$\delta p=c_a^2\delta\rho \tag{3.10}$$

We can, on the other hand, estimate the magnitude of the pressure fluctuation from the horizontal momentum equation; if

$$\nabla p=O\left(\rho\frac{\partial u}{\partial t}\right) \quad \text{then}$$

$$\delta p=O\left(\frac{\rho u \omega}{k}\right)$$

from which it follows from the relation between the pressure and density disturbances:

$$\frac{\delta\rho}{\rho}=O\left(\frac{u\omega}{kc_a^2}\right) \tag{3.11}$$

Thus,

$$\frac{1}{\rho}\frac{\partial \delta\rho}{\partial t}=O\left(\frac{u\omega^2}{c_a^2 k}\right)$$

We should compare this term, which is the estimate of the size of the local time derivative in the mass conservation equation with a typical term in the remaining combination of terms, namely, $\nabla \cdot \vec{u} = O(ku)$. Their ratio is thus

$$\frac{\dfrac{\partial \delta\rho}{\rho\partial t}}{\nabla \cdot \vec{u}}=O\left(\frac{u\omega^2}{kuk}\right)=\frac{\omega^2/k^2}{c_a^2}=\frac{c^2}{c_a^2} \qquad (3.12)$$

Thus, as long as the phase speed of the wave is small compared to the speed of sound, we can approximate the wave motion occurring as in an *incompressible fluid* for which the equation for mass conservation reduces to the condition

$$\nabla \cdot \vec{u} = 0 \qquad (3.13)$$

Note again that this *does not* by itself imply that $d\rho/dt = 0$. A separate consideration of the thermodynamics and the strength of the dissipation is required for that.

We now have a series of parameter tests we can make after the fact to check to see whether the approximations of

1. linear motion
2. inviscid motion
3. incompressible motion
4. nonrotating dynamics

will be valid.

Assuming that these conditions will be met by the waves under consideration here, the equations of motion reduce to the much simpler set:

$$\rho\frac{\partial \vec{u}}{\partial t}=-\nabla p - \rho g\hat{z} \qquad (3.14a)$$

$$\nabla \cdot \vec{u} = 0 \qquad (3.14b)$$

where $\hat{z}$ is a unit vector in the direction antiparallel to the direction of the local gravitation.

We could have just waved our hands (perhaps appropriately for a course on waves) and written down these traditional approximate equations. However, it is important for each new investigation of a wave type to carefully consider a priori the conditions required to achieve the approximate dynamics used for the physical description of the wave to make sure that our physical system is no more complicated than it need be, while at the same time, it should be consistent with the underlying physics of the fluid.

The curl of our momentum equation (recall that we are considering a fluid of constant density; the student is invited to use the thermodynamic equation to find the condition for the validity of that approximation) yields

$$\frac{\partial \nabla \times \vec{u}}{\partial t} = 0 \tag{3.15}$$

So, if the vorticity is zero initially or at any instant (as it would be for an oscillatory motion for which each field goes through zero periodically), it follows that it remains zero for all time. If the curl of the velocity is zero, it follows from a fundamental fact of vector calculus that the velocity can be represented by a *velocity potential*, ϕ,

$$\vec{u} = \nabla \phi \tag{3.16}$$

Note that only the *spatial* gradients of the velocity potential carry physical information. Any arbitrary function of time can be added to ϕ without changing the velocity field.

Since the motion is incompressible,

$$\nabla \cdot \vec{u} = \nabla \cdot (\nabla \phi) = \nabla^2 \phi = 0 \tag{3.17}$$

The equation of motion within the fluid thus reduces to the *elliptic* problem governed by Laplace's equation:

$$\nabla^2 \phi = 0 \tag{3.18}$$

This is an amazing simplification, and it should be a little disconcerting, because we are looking to describe a wave motion. Laplace's equation, by itself, is certainly not a wave equation. It describes among other things the electrical potential of *static* charges as well as certain static gravitational fields but, alone, no dynamical wave mechanism. The resolution of this seeming paradox is of course connected to the fact that we have not yet considered the boundary conditions for our problem. There is no more illuminating example of the importance of *boundary conditions* in the specification of the problem than this case of surface gravity waves. All the dynamics are in the boundary conditions. The internal equation, i.e., Laplace's equation merely relates the horizontal and vertical structure of the motion field.

Boundary Conditions

The obvious boundary condition at the lower horizontal surface is that the normal velocity vanishes there, i.e., $w = 0$ at $z = -D$, or

$$\frac{\partial \phi}{\partial z} = 0, \quad z = -D \tag{3.19}$$

The boundary conditions at the upper surface are significantly more interesting. Let's call the departure of the free surface from its level "rest position" $\eta(x, y, t)$ (Fig. 3.2),

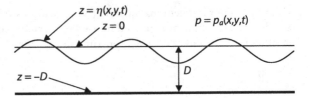

Fig. 3.2.
A definition figure for variables describing the motion in the surface gravity wave

which must be small (this will presently be made more explicit). Thus, we consider the rippled free surface to only be slightly in departure from its rest state.

At the free surface, the physical boundary conditions are

1. the dynamic condition:

$$p(x,y,z,t) = p_a(x,y,t) \quad z = \eta \tag{3.20}$$

and
2. the kinematic condition:

$$w = \frac{\partial \phi}{\partial z} = \frac{d\eta}{dt} \Rightarrow \frac{\partial \eta}{\partial t} \tag{3.21a}$$

or

$$\frac{\partial \phi}{\partial z} = \frac{\partial \eta}{\partial t}, \quad z = \eta \tag{3.21b}$$

We must now write these conditions completely in terms of the velocity potential, ϕ. The linearized momentum equation is

$$\frac{\partial \vec{u}}{\partial t} \equiv \frac{\partial \nabla \phi}{\partial t} = -\frac{\nabla p}{\rho} - g\nabla z \tag{3.22a}$$

or

$$\nabla \left\{ \frac{\partial \phi}{\partial t} + \frac{p}{\rho} + gz \right\} = 0 \tag{3.22b}$$

The integral of the last equation implies that

$$\frac{\partial \phi}{\partial t} + \frac{p}{\rho} + gz = F(t) \tag{3.23}$$

where $F(t)$ is an *arbitrary* function only of time. We can always add a function that is only of time to the velocity potential without changing the physical meaning of that potential. Let's imagine that we have added such an additional term such that its derivative with respect to time is equal to $F(t)$. This allows us to write this linearized form of Bernoulli's equation *everywhere* in the fluid in the form:

$$\frac{\partial \phi}{\partial t} + \frac{p}{\rho} + gz = 0 \tag{3.24}$$

Now let's apply this equation to the upper surface where $z = \eta(x,y,t)$ and $p = p_a(x,y,t)$. Thus,

$$\frac{\partial \phi}{\partial t} + \frac{p_a}{\rho} + g\eta = 0 \tag{3.25}$$

A derivative of this equation with respect to time yields, using the kinematic condition on the upper surface,

$$\frac{\partial^2 \phi}{\partial t^2} + g\frac{\partial \phi}{\partial z} = -\frac{1}{\rho}\frac{\partial p_a}{\partial t}, \quad \text{at} \quad z = \eta \tag{3.26}$$

Note that each term in this boundary condition is linear.

The condition is applied at the *unknown* location $z = \eta$. Indeed, the position of the free surface is, after all, one of the principal unknowns of the problem that we are trying to predict. For the general nonlinear problem, this unknown location of the boundary, at which the important boundary conditions are applied, is one of the most difficult aspects of the problem. However, we are considering only the linear small amplitude problem, and it turns out that we can apply the boundary condition at the *original* position of the interface, i.e., at $z = 0$. To see this take any term on the left-hand side of the above boundary condition, generically call it $G(x,y,\eta)$ and expand it around $\eta = 0$. Thus,

$$G(x, y, \eta) = G(x, y, 0) + \eta \frac{\partial G}{\partial z}\bigg|_{z=0} + \text{ higher order terms} \tag{3.27}$$

The first term on the right-hand side of the above equation is of the order of the amplitude of the motion, since G is one of the dynamic variables. Note that *all* the dynamical variables are linear in the size of the amplitude of the motion.

The second term is of the order of G times the free surface height and is therefore of the order of the amplitude squared. To be consistent with our linearization, such quadratic terms *must* be neglected. That implies that each term in the boundary condition stated above can be applied at $z = 0$; thus, we have

$$\frac{\partial^2 \phi}{\partial t^2} + g\frac{\partial \phi}{\partial z} = -\frac{1}{\rho}\frac{\partial p_a}{\partial t}, \quad \text{at} \quad z = 0 \tag{3.28}$$

as the boundary condition on the upper surface, while again at the lower surface,

$$\frac{\partial \phi}{\partial z} = 0, \quad z = -D \tag{3.29}$$

Note that the upper boundary condition contains two time derivatives. This is the mathematical source of the wave motion we will be describing. Its physical source is the interplay between the gravitational force at the upper boundary providing a restoring force and the relation between the free surface elevation and the vertical velocity at the upper surface.

We must also specify boundary conditions on the *lateral boundaries*. The simplest problem we will consider will be that of a wave in an infinitely broad layer of fluid. This is clearly an approximation, and we imagine that such a description will be valid until the waves to be found propagate and interact with the inevitable lateral boundaries of the fluid. Until that time, we may provisionally just insist that the solutions remain finite as x and y go to infinity. Useful references for formulation of the gravity wave problem can be found in Kundu (1990), Lamb (1945) and Stoker (1957).

Plane Wave Solutions for Surface Gravity Waves: Free Waves ($p_a = 0$)

In Cartesian coordinates, Laplace's equation can be written as

$$\frac{\partial^2 \phi}{\partial x^2} + \frac{\partial^2 \phi}{\partial y^2} + \frac{\partial^2 \phi}{\partial z^2} = 0 \tag{3.30}$$

We clearly can't have a *three*-dimensional plane wave because

1. the operator (the Laplacian) won't allow it, since an attempt to find such a plane wave would lead to the condition

$$k_1^2 + k_2^2 + k_3^2 = 0 \tag{3.31}$$

   which is impossible if all three components of the wave vector are real;
2. a plane wave won't satisfy the boundary condition $\partial \phi / \partial z = 0$ at $z = -D$;
3. the boundary conditions of finiteness as x and y get large imply that the horizontal components of the wave number are real, and thus the vertical component, k_3 or m, must be purely imaginary.

We can find solutions, however, that are periodic in x, y, and t of the form

$$\phi = R(z)e^{i(kx+ly-\omega t)} \tag{3.32}$$

where again, the real part of the above equation is meant. Substitution into Laplace's equation yields the ordinary differential equation for $R(z)$,

$$\frac{d^2 R}{dz^2} - K^2 R = 0 \tag{3.33a}$$

$$K^2 = k^2 + l^2 \tag{3.33b}$$

Note that all Laplace's equation will do is determine the structure with depth of the plane wave solution in x and y.

The solution for R that satisfies the kinematic boundary condition at $z = -D$, i.e., that $dR/dz = 0$, is

$$R = A \cosh K(z + D) \tag{3.34}$$

When this form is substituted into the boundary condition at $z = 0$, we obtain as a condition for a nonzero solution for A

$$-\omega^2 \cosh(KD) + gK \sinh(KD) = 0 \tag{3.35}$$

or

$$\omega = \pm\sqrt{gK \tanh KD} \tag{3.36}$$

and

$$c = \frac{\omega}{K} = \pm(gD)^{1/2} \left[\frac{\tanh KD}{KD} \right]^{1/2} \tag{3.37}$$

There are several important things to note about these results.

1. For each wave vector amplitude K, there are two waves propagating in opposite directions, parallel and antiparallel to the wave vector. The frequency and phase speed depend only on the wavelength, i.e., K and not on the orientation of the wave vector.
2. The phase speed is *different* for different wavelengths in distinction to light waves or sound waves. A pattern made out of a superposition of plane waves of different wavelengths will have each component move at a different speed and hence the pattern will *disperse*, which is why the relation between frequency and wave number is called the *dispersion relation*.
3. There are some important limiting cases to consider.

The maximum phase speed occurs when the wavelength (inverse to K) is long compared to the depth, i.e., when $KD \ll 1$. Then the phase speed approaches $(gD)^{1/2}$ and is independent of wavelength in that limit. In that case, when the phase speed is independent of wavelength, the wave is called *nondispersive*. In that long wave limit or for *shallow water waves*, $\omega = K(gD)^{1/2}$.

On the other hand, when the wavelength is short compared with the depth, i.e., when $KD \gg 1$, the dispersion relation becomes independent of depth and $\omega = (gK)^{1/2}$, while $c = (g/K)^{1/2}$. These *deepwater waves* are clearly dispersive. We will have to investigate why the frequency and phase speed become independent of D in this limit.

Now that we have the phase speed, we can check our assumption of incompressibility, that is, is $c \ll c_a$. Since the maximum phase speed is given by the shallow water limit for that condition, it will be satisfied if

$$\sqrt{gD} \ll c_a$$

or

$$D \ll c_a^2 / g$$

For water, the sound speed is of the order of $1\,400$ m s^{-1}. That places a condition on the depth such that for the incompressibility condition to be valid, we require $D \ll 200$ km (pretty safe for oceanography, at least on Earth).

The nature of the dispersion relation is evident in Fig. 3.3 showing the frequency, phase speed and group velocity as a function of wave number.

The dispersive nature of the waves can be contrasted to that of the standard "wave equation" (which we will see in this course captures only a small fraction of wave physics in oceanography and meteorology). For light waves in a vacuum, sound waves and waves on a thin string, the governing equation in one dimension is of the form

$$\frac{\partial^2 \varphi}{\partial t^2} - a^2 \frac{\partial^2 \varphi}{\partial x^2} = 0 \tag{3.38}$$

whose general solution is known to be

$$\varphi = F(x + at) + G(x - at) \tag{3.39}$$

consisting of two pulses traveling with the constant phase speeds $\pm a$. The forms F and G are determined by initial conditions after which the pulses travel without fur-

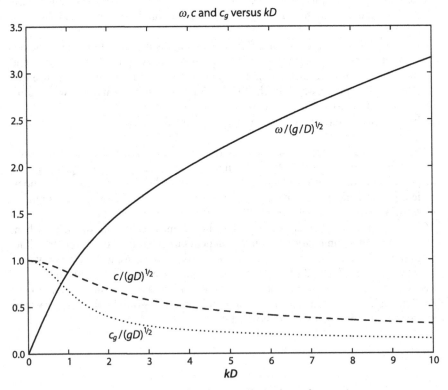

Fig. 3.3. Curves of frequency, phase speed and group velocity for surface gravity waves

ther change of shape. These are the classic nondispersive solutions for waves. In our case, the waves are highly dispersive and the evolution of the wave shape with time and unraveling the subsequent propagation of properties in the waves is a problem of great subtlety and interest. It will eventually, as we might imagine from our earlier discussion, come to depend on the character of the group velocity. For gravity waves, with the dispersion relation quoted above

$$\vec{c}_g = \nabla_{\vec{K}} \omega(K) = \frac{\partial \omega}{\partial K} \frac{\vec{K}}{K} \tag{3.40}$$

Thus, since the frequency is a function only of K, the group velocity is parallel to the wave vector and hence parallel to the direction of phase propagation.

With $\omega^2 = gK \tanh KD$,

$$2\omega \frac{\partial \omega}{\partial K} = g \left\{ \tanh KD + \frac{KD}{\cosh^2 KD} \right\} \tag{3.41a}$$

or

$$2cc_g = \frac{g}{K} \left\{ \tanh KD + \frac{KD}{\cosh^2 KD} \right\} \tag{3.41b}$$

and

$$\frac{c_g}{c} = \frac{1}{2} \left\{ 1 + \frac{2KD}{\sin 2KD} \right\} \tag{3.41c}$$

Thus the group velocity coincides with the phase speed for long waves ($KD \ll 1$), while for short waves the group velocity is 1/2 the phase speed (see Fig. 3.4).

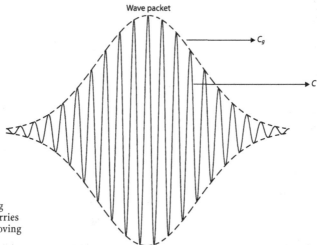

Fig. 3.4.
A wave packet propagating with the group velocity carries a plane wave with crest moving with the phase speed

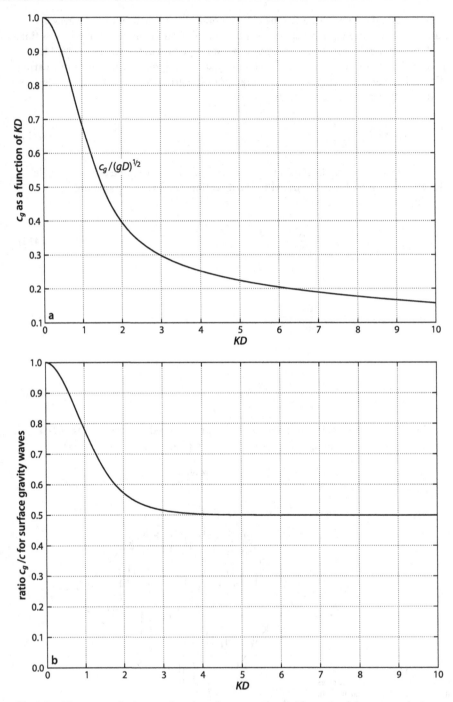

Fig. 3.5. a The group velocity as a function of wave number. **b** The *ratio* of the group velocity to phase speed for surface gravity waves as a function of wave number scaled with fluid depth, i.e., *KD*

Suppose we have a wave packet carrying a short wave, $KD \ll 1$.

The amplitude and K will move with the group velocity, while individual crests will move with the phase speed. Since c_g is half the phase speed for short waves, we will see individual crests appearing at the rear of the packet and travelling through the moving packet to disappear at the leading edge of the packet. Where do the crests go? Well, they are only a feature of the pattern, and they appear and disappear like smiles. It is the wave envelope moving with the group velocity that has physical content.

The ratio of the group velocity to the phase speed is shown in Fig. 3.5b as a function of wave number. They are equal for the longest waves, while for short waves the group velocity is *half* the phase speed.

Fields of Motion in Gravity Waves and Energy

Now that we have the dispersion relation, i.e., the dependence of frequency on *wave number* (we define the magnitude, K, of the wave vector K to be the wave number), we can ask what the fluid motion is in the wave field.

Our plane wave solution has been written in the form:

$$\phi = Ae^{i(\vec{K}\cdot\vec{x}-\omega t)}\cosh K(z+D) \tag{4.1}$$

Using the boundary condition at $z = 0$,

$$g\eta = -\frac{\partial \phi}{\partial t}\, z = 0 \tag{4.2}$$

since p_a has been taken to be zero for these *free waves*. We can therefore calculate the free surface elevation from Eq. 4.1 and Eq. 4.2,

$$\eta = \left(\frac{i\omega}{g}\right)Ae^{i(\vec{K}\cdot\vec{x}-\omega t)}\cosh KD$$

and it is understood the real part of each expression is to be taken.

A is an arbitrary amplitude, and it will be useful to consider the amplitude of the disturbance in terms of the amplitude of the free surface perturbation. So, let's define

$$\eta_0 = i\left(\frac{A\omega}{g}\right)\cosh KD$$

and take it to be real (this only defines the zero of the spatial phase, the point where the free surface elevation is a maximum). This yields

$$\eta = \eta_0 \cos(\vec{K}\cdot\vec{x}-\omega t) \tag{4.3a}$$

$$\phi = \eta_0 \frac{\omega}{K}\frac{\cosh K(z+D)}{\sinh KD}\sin(\vec{K}\cdot\vec{x}-\omega t) \tag{4.3b}$$

From the velocity potential, we can calculate each velocity component, since $\vec{u} = \nabla\phi$.

From the above formula for ϕ, we calculate the horizontal velocity vector and the vertical component of velocity:

$$\vec{u}_{\mathrm{H}} = \eta_0 \omega \left(\frac{\vec{K}}{K} \right) \cos(\vec{K} \cdot \vec{x} - \omega t) \frac{\cosh K(z+D)}{\sinh KD} \tag{4.4a}$$

$$w = \eta_0 \omega \sin(\vec{K} \cdot \vec{x} - \omega t) \frac{\sinh K(z+D)}{\sinh KD} = \frac{\partial \eta}{\partial t} \frac{\sinh K(z+D)}{\sinh KD} \tag{4.4b}$$

From the Bernoulli equation

$$p = -\rho g z - \rho \frac{\partial \phi}{\partial t}$$

we can calculate the pressure field in the wave. (Note that part of the pressure field has nothing to do with the wave. That is the first term on the right-hand side; it is present even in the absence of the disturbance). From the result from the velocity potential we obtain

$$p = -\rho g z + \rho g \eta_0 \cos(\vec{K} \cdot \vec{x} - \omega t) \frac{\cosh K(z+D)}{\cosh KD}$$

$$= \rho g \left[\eta \frac{\cosh K(z+D)}{\cosh KD} - z \right] \tag{4.5}$$

There are some very important qualitative features to note before moving on.

1. The horizontal velocity, $\vec{u}_{\mathrm{H}}$, is in the direction of the wave vector and hence in the direction of the propagation of the wave. This is not surprising for anyone who has lolled in the surf and felt himself move back and forth in the direction of a wave as it has passed by;

2. Each perturbation variable is proportional to the amplitude of the free surface elevation. That is, in this linear problem, the amplitude of every aspect of the motion is proportional to the free surface elevation. This implies that products of any two motion variables must be quadratic in the surface elevation (this is what we used to linearize the surface boundary condition);

3. In the limit of deep water or equivalently short waves for which $KD \gg 1$, the asymptotic forms of the hyperbolic functions imply that

$$\frac{\cosh K(z+D)}{\sinh KD} \approx \frac{e^{K(z+D)}}{e^{KD}} = e^{Kz}, \quad z < 0 \tag{4.6a}$$

$$\frac{\sinh K(z+D)}{\sinh KD} \approx \frac{e^{K(z+D)}}{e^{KD}} = e^{Kz} \tag{4.6b}$$

so that all the dynamical variables decrease exponentially from the free surface. The scale of decrease, as imposed by Laplace's equation, is just the wavelength. Hence, for waves whose wavelength is short compared to the depth, the motion decays long before the bottom is reached. The wave field then does not sense the presence of the bottom. This is why the frequency/wave number relation becomes independent of D as KD gets large. It's a good rule of thumb to remember for gravity waves; the depth of influence of the wave is its wavelength.

4. For very long waves, or equivalently, for shallow water, such that $KD \longrightarrow 0$, the limiting form of the hyperbolic functions yields

$$KD \ll 1$$

$$\vec{u}_H = \left(\frac{\eta}{D}\right) c \frac{\vec{K}}{K} \tag{4.7a}$$

$$w = \frac{\partial \eta}{\partial t}(z + D) \tag{4.7b}$$

$$p = \rho g(\eta - z) \tag{4.7c}$$

In this limit, the *horizontal velocity is independent of depth*. Its magnitude is the ratio of the free surface elevation to the depth multiplied by the phase speed. Thus, as long as $\eta / D \ll 1$, it will follow that $\vec{u}_H / c \ll 1$, which is the condition for linearization. The vertical velocity is proportional to the rate of displacement of the free surface, linearly diminishing to zero at the bottom, and the *pressure field is in hydrostatic balance in this limit*.

It is left to the student to show that in the short wave limit, $KD \ll 1$, the condition for linearization, $u \ll c$, leads directly to the condition $\eta_0 K \ll 1$. That is, the free surface displacement divided by the wavelength must be small, i.e. the slope of the free surface must be small.

Energy and Energy Propagation

The kinetic energy in a gravity wave per unit volume is simply

$$\frac{\rho |\vec{u}|^2}{2}$$

where the magnitude of the velocity is denoted by the vertical bars. Integrated over depth we have the *kinetic energy per unit horizontal area*,

$$KE = \int_{-D}^{0} dz \frac{\rho |\vec{u}|^2}{2} \tag{4.8}$$

The potential energy per unit horizontal area is

$$PE = \int_{-D}^{\eta} \rho g z\, dz = \frac{\rho g}{2}\left(\eta^2 - D^2\right) \tag{4.9}$$

Note that the term proportional to D^2 in the PE is an irrelevant constant. Note, too, that we have integrated to the free surface elevation η in the expression for PE but only to $z = 0$ in the expression for KE. The reason for this is that we are calculating the energy to the *second* order in the wave amplitude, and to do this for the PE we must include the free surface displacement. If we were to extend the integral for KE to include η in the upper limit, the correction to the expression for KE would be of $O(u^2\eta)$, i.e., of third order in the small wave amplitude and hence negligible. So the above integrals as stated are each of order amplitude squared.

Now let's try to develop an equation for the propagation of wave energy. We start from the governing equation, which is Laplace's equation for the velocity potential. An excellent discussion of wave energy and its propagation can also be found in Kundu (1990) and Stoker (1957).

We multiply that equation by the time derivative of the potential, viz.

$$
\begin{aligned}
0 = \frac{\partial \phi}{\partial t}\nabla^2\phi &= \nabla \cdot \left(\frac{\partial \phi}{\partial t}\nabla\phi\right) - \nabla\phi \cdot \frac{\partial \nabla\phi}{\partial t} \\
&= -\frac{\partial}{\partial t}\left[\frac{(\nabla\phi)^2}{2}\right] + \nabla \cdot \left(\frac{\partial \phi}{\partial t}\nabla\phi\right)
\end{aligned}
\tag{4.10}
$$

We recognize that the first term is (minus) the rate of change of kinetic energy per unit volume.

Let's now integrate the above equation over the depth of the fluid.

$$
-\frac{\partial}{\partial t}\int_{-D}^{0}dz(\nabla\phi)^2/2 + \int_{-D}^{0}\nabla_H \cdot \left(\nabla\phi\frac{\partial \phi}{\partial t}\right)dz + \int_{-D}^{0}\frac{\partial}{\partial z}\left(\frac{\partial \phi}{\partial z}\frac{\partial \phi}{\partial t}\right)dz = 0
\tag{4.11}
$$

Here the symbol ∇_H is the portion of the divergence in the horizontal plane, i.e.,

$$
\nabla_H \cdot \vec{Q} \equiv \frac{\partial Q_x}{\partial x} + \frac{\partial Q_y}{\partial y}
$$

where $\vec{Q}$ is any vector.

The last term in the equation above can be integrated, and the resulting terms evaluated using the boundary conditions. Since $\partial\phi/\partial z = 0$ at $z = -D$, while

$$
\frac{\partial \phi}{\partial z} = \frac{\partial \eta}{\partial t}
\tag{4.12a}
$$

$$
\frac{\partial \phi}{\partial t} = -g\eta, \quad z = 0
\tag{4.12b}
$$

we obtain

$$
-\frac{\partial}{\partial t}\int_{-D}^{0}dz(\nabla\phi)^2/2 + \int_{-D}^{0}\nabla_H \cdot \left(\nabla\phi\frac{\partial \phi}{\partial t}\right)dz - g\eta\frac{\partial \eta}{\partial t} = 0 \quad \text{or}
\tag{4.13a}
$$

$$
\frac{\partial}{\partial t}[KE + PE] + \nabla_H \cdot \vec{\Im} = 0
\tag{4.13b}
$$

$$
\vec{\Im} = -\int_{-D}^{0}\frac{\partial \phi}{\partial t}\nabla_H\phi\,dz
\tag{4.13c}
$$

That is, the rate of change, locally, of the total energy per unit horizontal area is balanced by the horizontal divergence of the *flux of wave energy*, $\vec{\Im}$, a horizontal vector. This horizontal flux can be easily interpreted physically, since

$$-\frac{\partial \phi}{\partial t} \nabla_H \phi = (p + \rho g z)\vec{u}_H \tag{4.14}$$

and $(p + \rho g z) = p'$, which is the part of the pressure field due to the wave activity. Therefore, the energy flux vector is just the rate at which the pressure field in the wave is doing work on the surrounding fluid. That rate of work yields the energy transfer from one part of the fluid to another and hence the energy flux. We shall often be looking for energy balance equations of the above type, i.e.,

$$\frac{\partial E}{\partial t} + \nabla \cdot \vec{\Im} = \text{sources} + \text{dissipation}$$

that is, the rate of change of wave energy locally and its flux to other parts of the fluid balanced by sources and sinks of energy. In the present case of a free, inviscid gravity wave, both the sources and sinks are zero.

An interesting question arises here. If, as we believe, the important physical attributes in the wave field propagate with the group velocity, can we relate the energy flux vector to the group velocity?

First, let us calculate the kinetic and potential energy in the field of motion of the plane wave we have been discussing. To make life easier for ourselves (always a good idea) let us orient our x-axis to coincide with the direction of the wave vector. Then, since the horizontal velocity is in the direction of the wave vector as shown above, there will be only the x-component of the horizontal velocity to deal with along, of course, with w. In this coordinate frame, $K = k$.

The potential energy is easy to calculate:

$$PE = \frac{\rho g \eta_0^2}{2} \cos^2(kx - \omega t) \tag{4.15}$$

This form oscillates between its maximum and zero during a wave period. The significant quantity for our purposes is the average over a wave period, denoted by brackets, i.e.,

$$\langle PE \rangle = \frac{\rho g \eta_0^2}{4} \tag{4.17}$$

For *KE* we have

$$KE = \int_{-D}^{0} dz \rho (u^2 + w^2)/2 = \int_{-D}^{0} \rho \eta_0^2 \omega^2 \left[\begin{array}{l} \cos^2(kx - \omega t)\dfrac{\cosh^2 k(z+D)}{\sinh^2 kD} \\[2mm] + \sin^2(kx - \omega t)\dfrac{\sinh^2 k(z+D)}{\sinh^2 kD} \end{array} \right] dz/2 \tag{4.18a}$$

and

$$\langle KE \rangle = \rho \eta_0^2 \omega^2 \int_{-D}^{0} \frac{\cosh 2k(z+D)}{4\sinh^2 kD} dz = \rho \eta_0^2 \omega^2 \frac{\sinh 2kD}{8k\sinh^2 kD}$$

(4.18b)

$$= \rho \eta_0^2 gk \tanh kD \frac{\sinh kD \cosh kD}{4\sinh^2 kD} = \frac{\rho g \eta_0^2}{4}$$

In deriving this result, we have first used the averaging of the cosine and sine terms over a wave period, then the identity relating the square of the cosh and sinh terms to cosh of twice the argument and then finally the dispersion relation itself to write ω^2 in terms of the wave number.

We note the important fact that averaged over a wave period (or a wavelength if we were to average in x instead of t), the kinetic and potential energies are equal; that is, there is *equipartition* of energy in the wave field between potential and kinetic energy exactly as in the oscillation of a pendulum.

The total energy averaged over a period is

$$\langle E \rangle = \langle KE \rangle + \langle PE \rangle = \frac{\rho g \eta_0^2}{2}$$

(4.19)

Now let's calculate the energy flux vector in the x-direction and its average over a period.

$$\Im_x = -\rho \int_{-D}^{0} \frac{\partial \phi}{\partial t} \frac{\partial \phi}{\partial x} dz = \int_{-D}^{0} \frac{\rho \eta_0^2 \omega^3}{k\sinh^2 kD} \cos^2(kx-\omega t)\cosh^2 k(z+D)dz$$

$$\langle \Im_x \rangle = \frac{\rho \eta_0^2 \omega^3}{2k\sinh^2 kD} \int_{-D}^{0} \left[\frac{1}{2} + \frac{\cosh 2k(z+D)}{2} \right] dz$$

$$= \frac{\rho \eta_0^2 \omega^3}{2k\sinh^2 kD} \left[\frac{D}{2} + \frac{\sinh 2kD}{4k} \right]$$

$$= \frac{\rho \eta_0^2 ckD}{2\sinh^2 kD} g \tanh kD \left[\frac{1}{2} + \frac{\sinh 2kD}{4kD} \right]$$

$$= \frac{\rho g \eta_0^2 ck(D/2)}{\sinh kD \cosh kD} \left[\frac{1}{2} + \frac{\sinh 2kD}{4kD} \right] \quad (\text{note that } 2\sinh kD \cosh kD = \sinh 2kD)$$

$$= \rho g \eta_0^2 c \left[\frac{1}{4} + \frac{kD}{2\sinh 2kD} \right]$$

$$= \frac{\rho g \eta_0^2}{2} c \left[\frac{1}{2} + \frac{kD}{\sinh 2kD} \right]$$

$$= \frac{\rho g \eta_0^2}{2} c_g$$

$$= \langle c_g E \rangle$$

(4.20)

The important result obtained here is that for a plane gravity wave, the *horizontal flux of energy is equal to the energy itself multiplied by the group velocity.* That is equivalent to saying that the energy in the wave *propagates with the group velocity.* That is, the energy equation may be written:

$$\frac{\partial \langle E \rangle}{\partial t} + \nabla_{\mathrm{H}} \cdot \vec{c}_{\mathrm{g}} \langle E \rangle = 0 \tag{4.21}$$

In a uniform medium where the frequency and wave number are essentially constant, the group velocity will be independent of position. Thus for a wave packet, whose averaged energy just depends on the distribution of its envelope of free surface height, the above equation can be rewritten

$$\frac{\partial \langle E \rangle}{\partial t} + \vec{c}_{\mathrm{g}} \cdot \nabla \langle E \rangle = 0 \tag{4.22}$$

which states that for an observer moving laterally with the group velocity, the energy averaged over one phase of the wave is constant. *The energy in a slowly varying packet travels with the group velocity in a homogeneous medium.*

We will generalize this result to cases in which the energy is not simply contained in a compact packet, and we will see that the generalization also allows us to think of sequences of energy packets, each propagating with a group velocity appropriate for the wave number of that particular packet, which together with its companions represents an arbitrary disturbance.

Addendum to Lecture

With the velocity field given by the velocity potential, we can calculate the *trajectories* of fluid elements in the plane wave. Let ξ and ζ be the x and z *displacements* of the fluid elements around some original position (x_0, z_0). Then if the displacements are small, we can linearize the Lagrangian trajectory equations:

$$\frac{d\xi}{dt} = u(x_0 + \xi, z_0 + \zeta, t) \approx u(x_0, z_0, t) = \omega \eta_0 \cos(kx - \omega t) \frac{\cosh k(z + D)}{\sinh kD} \tag{4.23a}$$

and similarly

$$\frac{d\zeta}{dt} = w(x_0, z_0, t) = \omega \eta_0 \sin(kx - \omega t) \frac{\sinh k(z + D)}{\sinh kD} \tag{4.23b}$$

Integration yields

$$\xi = -\eta_0 \sin(kx - \omega t) \frac{\cosh k(z + D)}{\sinh kD} \tag{4.24a}$$

$$\zeta = \eta_0 \cos(kx - \omega t) \frac{\sinh k(z + D)}{\sinh kD} \tag{4.24b}$$

It follows that the trajectories are ellipses, i.e.,

$$\frac{\xi^2}{L_x^2} + \frac{\zeta^2}{L_z^2} = 1 \tag{4.25a}$$

$$L_x = \eta_0 \frac{\cosh k(z+D)}{\sinh kD} \tag{4.25b}$$

$$L_z = \eta_0 \frac{\sinh k(z+D)}{\sinh kD} \tag{4.25c}$$

Thus, the orbits are flat at the bottom of the fluid layer where $L_z = 0$. For deep water, the two axes of the ellipse are equal ($\eta_0 e^{kz}$), so the orbits are circularly shrinking in radius as z becomes more negative. For shallow water, the orbits reduce to essentially horizontal lines parallel to the bottom. The student is asked to discuss the direction of motion along the ellipse as the wave passes overhead.

The Initial Value Problem

It is not easy to see how a uniform or nearly uniform wave train can realistically emerge from some general initial condition or from a realistic forcing unless the initial condition or the forcing is periodic. That turns out not to be the case, and the ideas we have so far developed about group velocity and energy propagation turn out to be invaluable in getting to the heart of the general question of wave signal propagation. Indeed, it is the very dispersive nature of the wave physics (i.e., the dependence of the phase speed on the wave number) that is responsible for the emergence of locally nearly periodic solutions. This can be seen by examining the solution to the general initial value problem. This was first done by Cauchy in 1816. It was also solved at the same time by Poisson. The problem was considered so difficult at that time that the solution was in response to a prize offering of the Paris Académie (French Academy of Sciences). Now it is a classroom exercise.

We will again consider a disturbance that is a function only of x and z (and t of course), and we will consider the problem unforced by a surface pressure term, i.e., $p_a = 0$.

The layer is again of depth D and it is initially at rest.

As initial conditions, we will take

$$\eta(x,0) = N(x) \tag{5.1a}$$

$$\bar{u}(x,z,0) = 0 \;\Rightarrow\; \phi(x,z,0) = 0 \tag{5.1b}$$

The governing equation for the velocity potential is Laplace's equation, which for two dimensions is

$$\frac{\partial^2 \phi}{\partial x^2} + \frac{\partial^2 \phi}{\partial y^2} = 0 \tag{5.2}$$

with boundary conditions:

$$\frac{\partial \phi}{\partial z} = 0, \quad z = -D \tag{5.3a}$$

$$\left.\begin{aligned} w &= \frac{\partial \phi}{\partial z} = \frac{\partial \eta}{\partial t} \\ \frac{\partial \phi}{\partial t} &+ g\eta = 0 \end{aligned}\right\} z = 0 \Rightarrow \frac{\partial^2 \phi}{\partial t^2} + g\frac{\partial \phi}{\partial z} = 0, \quad z = 0 \tag{5.3b,c}$$

Since the region is infinitely long in the x-direction (in our approximation of a broad swath of open water) and the coefficients of the differential equations and boundary conditions are independent of x, it is appropriate and useful to represent the solution as a *Fourier Integral*. You may want to brush up on the Fourier integral by looking at any one of number of standard mathematical texts, e.g. Morse and Feshbach (1953). Thus, we write the velocity potential as

$$\phi(x,z,t) = \frac{1}{\sqrt{2\pi}} \int_{-\infty}^{\infty} \Phi(k,z,t)e^{ikx}dk \tag{5.4a}$$

with the dual return relation:

$$\Phi(k,z,t) = \frac{1}{\sqrt{2\pi}} \int_{-\infty}^{\infty} \phi(x,z,t)e^{-ikx}dx \tag{5.4b}$$

Note that the placement of the factors $\sqrt{2\pi}$ is somewhat arbitrary, and different conventions are used. The only requirement is that the product of the constant before each integral multiplies to $\frac{1}{2}\pi$.

Similarly for the free surface elevation,

$$\eta(x,t) = \frac{1}{\sqrt{2\pi}} \int_{-\infty}^{\infty} N(k,t)e^{ikx}dk \tag{5.5a}$$

$$N(k,t) = \frac{1}{\sqrt{2\pi}} \int_{-\infty}^{\infty} \eta(k,t)e^{-ikx}dk \tag{5.5b}$$

What we are doing is representing an arbitrary disturbance by an infinite sum of plane waves in x, whose wave numbers are a continuous distribution over all k, which is why an integral is required for the representation.

If the above representation for the potential is put into Laplace's equation, we obtain as a condition for the solution that at each wave number k,

$$\frac{\partial^2 \Phi}{\partial z^2} - k^2\Phi = 0 \tag{5.6}$$

while the boundary conditions become

$$\frac{\partial^2 \Phi}{\partial t^2} + g\frac{\partial \Phi}{\partial z} = 0, \quad z = 0 \tag{5.7a}$$

$$\frac{\partial \Phi}{\partial z} = 0, \quad z = -D \tag{5.7b}$$

and the similarity to the plane wave problem should be apparent. Indeed, the solution for Φ can be written:

$$\Phi(k,z,t) = A(k,t)\cosh k(z + D) / \sinh kD \tag{5.8}$$

This satisfies the boundary condition at $z = -D$. Satisfying the boundary condition on $z = 0$ requires

$$\frac{d^2A}{dt^2} + \omega(k)^2 A = 0 \tag{5.9}$$

where

$$\omega(k)^2 = gk \tanh kD \tag{5.10}$$

Thus, we can write

$$A(t) = a(k)e^{i\omega(k)t} + b(k)e^{-i\omega(k)t} \tag{5.11}$$

so that

$$\Phi = (ae^{i\omega t} + be^{-i\omega t})\frac{\cosh k(z+D)}{\sinh kD} \tag{5.12a}$$

$$\Rightarrow$$

$$\phi = \frac{1}{\sqrt{2\pi}} \int_{-\infty}^{\infty} \left[(ae^{i\omega t} + be^{-i\omega t})\right]e^{ikx}\frac{\cosh k(z+D)}{\sinh kD}dk \tag{5.12b}$$

The solution for the velocity potential consists of a sum of waves. For each k, one is moving to the left (the first term in square brackets) and the other is moving to the right (the second term). Each one is moving with the frequency associated with the plane wave at that k and with the vertical structure function of the plane wave at that k. The total solution is the integral sum of *all* the plane waves excited by the initial conditions.

Since

$$g\eta = -\frac{\partial \phi}{\partial t}\bigg)_{z=0} \tag{5.13a}$$

$$\eta(x,t) = -\frac{1}{g\sqrt{2\pi}} \int_{-\infty}^{\infty} \left[ae^{i\omega t} - be^{-i\omega t}\right]i\omega e^{ikx}\coth kD dk \tag{5.13b}$$

at $t = 0$ the velocity potential and its derivatives vanish. Thus for all k,

$$a(k) = -b(k) \tag{5.14}$$

and using the dispersion relation $\omega^2 = gk \tanh kD$,

$$\eta = \frac{1}{\sqrt{2\pi}} \int_{-\infty}^{\infty} \frac{ib(k)kdk}{\omega}\left[e^{i\omega t} + e^{-i\omega t}\right] \tag{5.15}$$

at $t = 0$

$$\eta = \eta_0(x) = \frac{1}{\sqrt{2\pi}} \int_{-\infty}^{\infty} N_0(k)e^{ikx}dk \tag{5.16}$$

which implies that

$$b(k) = \frac{N_0(k)\omega}{2ik} \tag{5.17}$$

or

$$\eta(x,t) = \frac{1}{\sqrt{2\pi}} \int_{-\infty}^{\infty} \frac{N_0(k)}{2} \left[e^{i\omega t} + e^{-i\omega t} \right] e^{ikx} dk \tag{5.18}$$

which has a simple interpretation, namely, that *half* of the initial condition at each k propagates to the left and the other half propagates to the right, each with the phase speed, frequency, and wave number relation of the plane wave of that k. We might have written down the above equation directly from our knowledge of the plane wave physics, but it is useful to go through the formal derivation at least once. Incidentally, now that $b(k)$ is known,

$$\phi = i \frac{1}{\sqrt{2\pi}} \int_{-\infty}^{\infty} \frac{N_0}{2} \left[e^{i\omega t} - e^{-i\omega t} \right] e^{ikx} \frac{\cosh k(z+D)}{\sinh kD} \tag{5.19}$$

This yields the *formal* solution to the problem, but it doesn't take much to realize that a solution written as an infinite integral is not very revealing, and our real work in understanding the physical nature of the initial value problem has just begun.

But first, to simplify things, let's assume that the initial condition on the free surface height is an even function of x around the origin, namely, $\eta_0(x) = \eta_0(-x)$. It follows from this that the Fourier transform of the initial condition, $N_0(k)$ is an even function of k. To show this,

$$N_0(k) = \frac{1}{\sqrt{2\pi}} \int_{-\infty}^{\infty} e^{-ikx} \eta(x) dx \quad \text{let } x = -\xi, \quad \text{then} \tag{5.20a}$$

$$N_0(k) = \frac{1}{\sqrt{2\pi}} \int_{-\infty}^{\infty} e^{ik\xi} \eta(-\xi) d\xi = \frac{1}{\sqrt{2\pi}} \int_{-\infty}^{\infty} e^{-i(-k)\xi} \eta(\xi) d\xi$$

$$= N_0(-k) \tag{5.20b}$$

where in the last step we have used the evenness of $\eta(x)$. Since $N_0(k)$ is an even function of k,

$$\eta = \frac{1}{\sqrt{2\pi}} \int_{-\infty}^{\infty} N_0(k) \cos \omega t e^{ikx} dk$$

$$= \frac{1}{\sqrt{2\pi}} \int_{-\infty}^{\infty} N_0(k) \cos \omega t \cos kx dk$$

$$= \sqrt{2/\pi} \int_{0}^{\infty} N_0(k) \cos \omega t \cos kx dk \tag{5.21}$$

Thus, we have succeeded in reducing the interval to the range $(0,\infty)$ in our k integration. Using a well-known identity for the product of cosine functions,

$$\eta = \sqrt{\frac{2}{\pi}} \int_0^\infty N_0(k)[\cos(kx+\omega t)+\cos(kx-\omega t)]\mathrm{d}k$$

$$= \mathrm{Re}\sqrt{\frac{2}{\pi}} \int_0^\infty N_0(k)\Big[e^{i(kx+\omega t)} + e^{i(kx-\omega t)}\Big]\mathrm{d}k \tag{5.22}$$

where again we recall that $\omega = (gk\tanh kD)^{1/2}$ (here we can take the positive root since we have explicitly included both signs of the solution in the above formulae).

At this point, we are still in the position of having our solution given in terms of an infinite integral. What can we say about the solution? Will some useful approximation teach us anything?

For short times, i.e., for a very small t, we could expand the expression for η as a power series in t, the first term of which is the known initial condition. That can allow us to examine the initial evolution of the disturbance. We might rightly object, saying that the initial evolution will depend very heavily on the arbitrary form of the initial condition. It will be much more illuminating to ask about the solution after a long time has passed so that the wave field can evolve to a state that reflects the general properties of the gravity wave field. Can we say something more useful, then? It turns out we can, using a classical method of approximating integrals of the type we have above: *the method of stationary phase.*

Our integrals for the free surface height are of the form

$$\eta = \sqrt{\frac{2}{\pi}} \int_0^\infty \frac{N_0(k)}{2}\Big[e^{it\chi(k)} + e^{it\psi(k)}\Big]\mathrm{d}k \tag{5.23a}$$

$$\chi(k) \equiv k(x/t) + \omega(k)t \tag{5.23b}$$

$$\psi(k) = k(x/t) - \omega(k)t \tag{5.23c}$$

and we would like to evaluate the integrals above for a large t and *with the ratio x/t fixed.* This is equivalent to saying that for a large t, we are evaluating the integrals moving away from the origin at the speed (arbitrary) $U = x/t$. So, for a large t, an arbitrary x should be chosen, which is also large. That determines $U = x/t$, and we want to find the value of the integral at that time and at that point.

The disturbance for $x > 0$ will be given by the second term in the above integral, so consider the second integral in the equation for η. Suppose that the function $\psi(k)$ does not vanish on the semi-infinite k interval. Then we could change the dependent variable of the integral from k to ψ, and obtain

$$\eta = \sqrt{\frac{2}{\pi}} \int_0^\infty \frac{N_0}{2(\mathrm{d}\psi/\mathrm{d}k)}e^{it\psi}\mathrm{d}\psi \tag{5.24}$$

Integration by parts yields

$$\eta = \frac{1}{it}\frac{N_0(k)}{\mathrm{d}\psi/\mathrm{d}k}e^{it\psi}\Big|_0^\infty - \frac{1}{it}\int_0^\infty e^{it\psi}\mathrm{d}\psi\frac{\mathrm{d}}{\mathrm{d}\psi}\frac{N}{\mathrm{d}\psi/\mathrm{d}k} \tag{5.25}$$

so that the disturbance would decay at least as fast as $1/t$ (in fact it will decrease much more rapidly, exponentially. See Lighthill, *Waves in Fluids* (1978) Jeffreys and Jeffreys (1962) or Stoker (1957).

This rapid decay with time is due to the fact that while $N_0(k)$ is a smooth function of k, the sinusoidal behavior of the exponential produces a factor that oscillates very rapidly when t is large as a *function of k*, so that contributions to the integral from some interval in k are cancelled at $k + \Delta k$ by a factor of the opposite sign, as shown in Fig. 5.1.

Thus, as long as $\psi(k)$ increases smoothly with k, the factor $e^{it\psi(k)}$ will oscillate very rapidly as a function of k for a large t, *unless in the neighborhood of some point k_s, the function $\psi(k)$ does not increase with k, i.e., unless that point is a stationary point at which*

$$\frac{d\psi}{dk}(k_s) = 0 \tag{5.26}$$

At such points, the phase function ψ will not increase with k, and there is an opportunity for the integral to accumulate value in that neighborhood.

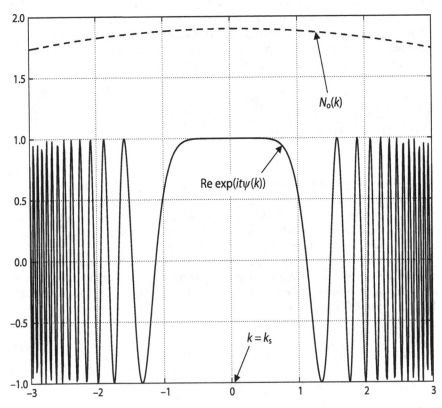

Fig. 5.1. The behavior of the exponential factor for a large t showing the interval of stationary phase

To find such points of *stationary phase*:

$$\psi = kU - \omega(k), \quad U = x/t \tag{5.27a}$$

$$\frac{d\psi}{dk} = 0 = U - \partial\omega/\partial k = x/t - c_g(k) \tag{5.27b}$$

Thus at a given x and t, or for an observer moving away from the origin at a speed x/t, the wave number of stationary phase, k_s is given by that wave number whose group velocity matches the velocity $U = x/t$ (Fig. 5.2).

We note that for a given x/t, a stationary phase wave number can be found as long as x/t is less than the maximum value of c_g in the whole k interval. Since the maximum value of the group velocity occurs for the longest wave and this maximum is $\sqrt{gD}$, we anticipate that for time t, the disturbance will be limited to a region $x \le t\sqrt{gD}$. Thus, there will be a front moving out from the origin at the speed $\sqrt{gD}$, ahead of which the fluid will be essentially undisturbed and behind which the solution will be given by the asymptotic approximation to the integral we will now develop (Fig. 5.3).

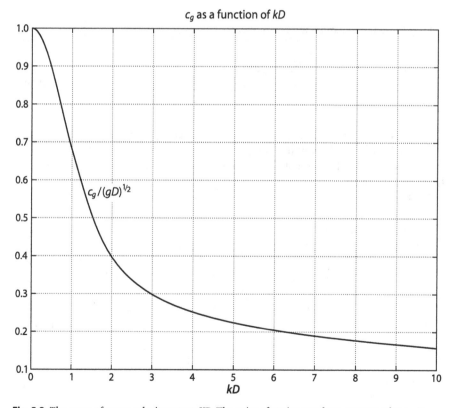

Fig. 5.2. The curve of group velocity versus KD. The point of stationary phase corresponds to $x/t = c_g$

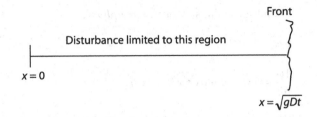

Fig. 5.3.
The interval for which the
disturbance can be found
for a large t

Consider the integral:

$$I = \int_0^\infty e^{it\psi} \frac{N_0(k)}{2} dk \qquad (5.28)$$

As we have argued, for a large t, the major contribution from this integral comes from the interval in k near the stationary point k_s. Near that point, we can write

$$\psi(k) = \psi(k_s) + \underbrace{\frac{d\psi}{dk}(k_s)}_{= 0 \text{ by defn.}} \cdot (k - k_s) + \frac{d^2\psi}{dk^2}(k_s) \cdot (k - k_s)^2 + \cdots \qquad (5.29)$$

Thus the integral can be approximated as

$$I \approx \int_0^\infty e^{it\psi(k_s)} e^{it\psi''(k_s)(k-k_s)^2/2} N_0(k_s) dk/2 \qquad (5.30)$$

$$\psi'' \equiv d^2\psi/dk^2$$

Note that we have replaced $N_0(k)$ by its value at the stationary point. This is valid since only in this vicinity will the integral have an asymptotic value greater than $1/t$ and N_0 is assumed to be a smooth function of k and hence much more slowly varying than $t\psi$ for a large t.

Thus,

$$I \approx \frac{N_0(k_s)}{2} e^{it\psi(k_s)} \int_0^\infty e^{it\psi''(k_s)(k-k_s)^2/2} dk \qquad (5.31)$$

where the integral really extends over a region centered on k_s.

Let

$$|\psi''(k_s)| \frac{(k-k_s)^2}{2} t = \vartheta^2 \qquad (5.33a)$$

or

$$k - k_s = \vartheta \left(\frac{2}{|\psi''(k_s)|t} \right)^{1/2} \qquad (5.33b)$$

This allows the integral to be written:

$$I \approx \frac{N_0(k_s)}{2} \frac{e^{it\psi(k_s)}}{\left\{ \frac{t|\psi''(k_s)|}{2} \right\}^{1/2}} \int_{-\infty}^{\infty} e^{i\vartheta^2 \, \mathrm{sgn}\,\psi''(k_s)} d\vartheta \tag{5.34}$$

where the extension of the limits to plus and minus infinity follows from the relation between k and θ for a large t. The remaining integral is a standard one and can be found in almost all integral tables:

$$\int_{-\infty}^{\infty} e^{i\vartheta^2 \, \mathrm{sgn}\,\psi''(k_s)} d\vartheta = \sqrt{\pi}\, e^{i(\pi/4)\mathrm{sgn}\,\psi''(k_s)} \tag{5.35}$$

Putting these results together leads us to our final formula for the asymptotic solution for the initial value problem for $x > 0$ and for a large t:

$$\eta \approx \frac{N_0(k_s)}{\left|\psi''(k_s)t\right|^{1/2}} e^{i(t\psi(k_s)+[\pi/4]\mathrm{sgn}\,\psi''(k_s))} \tag{5.36a}$$

$$\psi = \frac{x}{t}k - \omega(k) \tag{5.36b}$$

Discussion

Now let's try to interpret the solution, valid for a large x and t, shown in the boxed equation above.

We can think of the solution in the vicinity of the point (x,t) as a plane wave with amplitude:

$$A = \frac{N_0(k_s)}{\sqrt{t\psi''(k_s)}} \tag{5.37}$$

and a phase

$$\theta(x,t) = t\psi = k_s x - \omega(k_s)t \tag{5.38}$$

Notice that since the wave number k_s is a function of x and t through the stationary phase condition

$$\frac{\partial \omega}{\partial k}(k_s) = x/t$$

the dependence of the phase of x and t can be rather complicated.

However, consider our generalized definition of wave number:

$$\frac{\partial \theta}{\partial x} = k_s + x\frac{\partial k_s}{\partial x} - \frac{\partial \omega}{\partial k}t\frac{\partial k_s}{\partial x}$$

$$= k_s + \frac{\partial k_s}{\partial x}\left[x - c_g(k_s)t\right] \tag{5.39}$$

Hence, the local variation of phase in x is equal to k_s for $x/t = c_g(k_s)$, i.e., for an observer moving away from the origin of the disturbance with the group velocity associated with that wave number. Further, moving at that constant speed, the wave number remains constant if the observer moves with that group velocity. Similarly,

$$-\frac{\partial \theta}{\partial t} = \omega(k_s) - \frac{\partial k_s}{\partial t}\left[x - c_g(k_s)t\right] \tag{5.40}$$

so that the frequency will be equal to $\omega(k_s)$ for an observer moving at the group speed at the stationary wave number.

Thus, at some point in the wave train, the disturbance will look like a plane wave with the wave number and frequency (ω_s, k_s) related by the dispersion relation, and these local parameters will remain constant as the point moves away from the origin with the group speed. In other words, *the wave number and frequency propagate with the group speed even though the original disturbance need not be anything close to a periodic form.* This is a result valid for a large t. What has happened is that the spectrum of the disturbance sorts itself out wave number by wave number such that the part of the disturbance with wave number k_s propagates with its group velocity to the position $x = c_g t$. This happens for each k. The part of the spectrum with the fastest group velocity (the long waves in this case) will be found out in front and the slow waves will bring up the rear. This explains why, although the initial disturbance may be quite different from a plane wave (e.g., a gaussian in x), the disturbance with time can be locally approximated by a plane wave, justifying our earlier concentration on the properties of plane and nearly-plane waves. It is the dependence of the phase speed and group speed on k that *disperses* the original signal into a parade of local plane wave perturbations.

At any fixed x, the wave number will change with time as slower, shorter waves arrive at that x.

Again, let's consider the phase

$$\theta = t\psi = k_s x - \omega(k_s)t = k_s t\left[\frac{x}{t} - \frac{\omega(k_s)}{k_s}\right] = k_s t[x/t - c(k_s)] \tag{5.41}$$

Thus, if we move in such a way as to keep the wave number constant, $x/t = c_g(k_s) \neq c(k_s)$, then the phase will change for the observer. Such an observer will see individual crests and troughs moving past at a rate that depends on the difference between the group and phase velocities. If, on the other hand, one wishes to follow an individual crest, so that we set $x/t = c(k)$, then from our above results,

$$k = \frac{\partial \theta}{\partial x} = k_s + \frac{\partial k_s}{\partial x} t \left[c - c_g \right] \qquad (5.42)$$

so that following an individual crest implies that the wavelength associated with that crest will be changing with time. In this physics, you can always ride the same horse if you want, but it may be repeatedly changing size if you insist on staying on the same horse. In order to ride a horse that is always the same size, you will need to constantly change horses (crests).

We can work this out analytically and explicitly in the limit when the water is so deep that for all wave numbers possessing any reasonable amount of energy in the spectrum, $kD \gg 1$. In this limit,

$$\omega = (gk)^{1/2} \qquad (5.43a)$$

$$c_g = \frac{1}{2} \left(\frac{g}{k} \right)^{1/2} = \frac{c}{2} \qquad (5.43b)$$

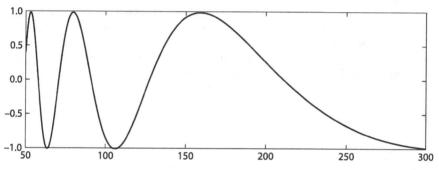

a Free surface as function of x, $t = 10$

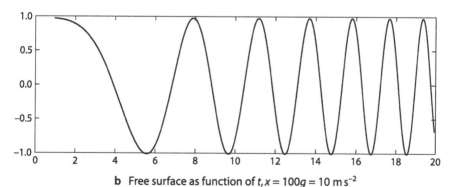

b Free surface as function of t, $x = 100g = 10 \text{ m s}^{-2}$

Fig. 5.4. The upper panel (**a**) shows the free surface at a fixed time. Note the long waves out in front. The bottom panel (**b**) shows the surface height field as a function of time at a fixed point. The low frequency waves (small k) arrive first and the higher frequencies arrive later, since they have slower group velocities

To find what the wave number of stationary phase is at the point x at time t,

$$c_g = x/t = \frac{1}{2}\left(\frac{g}{k_s}\right)^{1/2} \tag{5.44a}$$

$$k_s = \frac{1}{4}\frac{gt^2}{x^2} \Rightarrow \omega(k_s) = \frac{1}{2}gt/x \tag{5.44b}$$

$$\theta(k_s) = k_s x - (gk_s)^{1/2}t = -\frac{1}{4}gt^2/x \tag{5.44c}$$

$$c(k_s) = 2x/t \tag{5.44d}$$

Note that at a *fixed* position, the wave number increases (waves get shorter) with time, while the waves with the slower group velocities arrive later. At any given time, the waves get longer (k gets smaller) as x increases. Note that the phase at any x and t will change with time according to the ratio gt^2/x. To ride a particular crest, an observer must then move so that $x \propto t^2$, that is, the observer must *accelerate with time to keep up with a particular phase*. To follow a particular wave number, the observer *must move at a constant speed equal to the group velocity for that wave number*. Hence, for dispersive waves, one can not simultaneously keep to the same phase and the same wavelength, since the phase speeds and group velocities are not the same (Fig. 5.4).

Discussion of Initial Value Problem (*Continued*)

We have seen that the initial spectrum of the waves, which is initially localized in space, gets strung out with time so that at time t, each wave number appears at $x = c_g(k)t$. We might expect that the energy, if conserved, would also be distributed by wave number, so that the amount of energy at wave number k in the original spectrum at wave number k would also be found at the position $x = c_g(k)t$ for a large enough time. This is as if the original disturbance is composed of an infinite number of packets of constant wave number, each of which moves away from the origin of the disturbance with its own group velocity. Each satchel of energy moves with the group velocity (Fig. 6.1).

Let's try to make this more quantitative, and we will at the same time be able to explain the inverse dependence of the amplitude on the square root of time found in the last lecture. The energy in the gravity wave field is, as we have seen, proportional to the square of the free surface displacement. By a fundamental theorem of Fourier analysis,

$$E = \frac{\rho g}{2} \int_{-\infty}^{\infty} \eta^2 dx = \frac{\rho g}{2} \int_{-\infty}^{\infty} |N(k)|^2 dk \qquad (6.1)$$

which only states that we can count the energy in space or with the wave number spectrum.

Now, following an argument due originally to Rossby (1945) consider the energy in a spatial interval between x_s and $x_s + \Delta x_s$ such that the center of this infinitesimal interval is the place where the wave number k_s is found at time t.

Again, $x = c_g(k_s)t$.

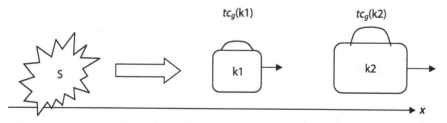

Fig. 6.1. A disturbance initiated by an initial source of energy, S, propagates away and is distributed among "suitcases" of energy, each moving with its group velocity

The x interval Δx_s will, for long times, be related to a wave number interval in the original spectrum by the relation

$$\Delta x_s = t \frac{\partial c_g}{\partial k}\bigg)_{k=k_s} \quad \Delta k_s = t \frac{\partial^2 \omega}{\partial k^2} \Delta k_s \tag{6.2}$$

The energy in that spatial interval from our asymptotic formula for the wave height will be (assuming N_0 is an even and real function)

$$\eta^2 \Delta x = \frac{N_0(k_s)^2}{\left(t\left|\dfrac{\partial^2 \omega}{\partial k^2}(k_s)\right|\right)^{1/2}} \cos^2(\theta(k_s) + i\,\mathrm{sgn}(\omega''(k_s)\pi/4)\Delta x_s \tag{6.3}$$

Averaging over a period and using the above expression for the interval length,

$$\eta^2 \Delta x_s = \frac{N_0(k_s)^2}{2} \Delta k_s \quad \text{(half goes the other direction)} \tag{6.4}$$

The above expression is a function only of k_s and so will remain constant for an observer moving at the group velocity. Thus, the energy in the original spectrum in the wave number interval Δk is conserved as it propagates outward with the group velocity. The length of the interval that energy is contained in continuously and linearly extends with time, because the group velocity is slightly different at the leading and trailing edges of the interval, since k is a continuous function of x at a given time. In order to have the energy conserved, the product of the amplitude squared times the interval length must be constant. Since the latter increase linearly with t, the amplitude *must decrease like* $t^{-1/2}$ to conserve energy. This explains the square root factor in the result of the previous lecture. Note that the contribution to the wave amplitude for those parts of the integral not near the stationary phase point will decline at least as fast as $1/t$. Then as time goes on, the stationary phase contribution will become increasingly dominant.

In the sense described above, the energy propagates with the group velocity. That is, energy present in the original spectrum at a given k finds itself at a position consistent with the group velocity as the propagation speed for energy.

Looking carefully at the result for the amplitude, we note that there is a potential difficulty with the expression for those values of k corresponding to the maximum (or minimum, should one exist) of the group velocity. At such values of k,

$$\frac{\partial^2 \omega}{\partial k^2} = \frac{\partial c_g}{\partial k} = 0 \tag{6.5}$$

and a singularity occurs. This, of course, coincides with a particularly interesting position in the wave train corresponding to (in the case of the maximum) the very front of the wave train where the fastest group can be found. Since the second derivative of frequency with respect to k vanishes at that k (this would be $k = 0$ for the gravity wave case), the expansion of ω as a function of k around k_s must be carried to a higher order, i.e., to order $(k - k_s)^3$. A discussion of the asymptotics can be found in many texts (e.g., Whitham 1974 or Stoker 1957). As one might imagine, since the group velocity is changing much more slowly where the derivative of the group velocity is nearly zero, the x-interval spreads more slowly, and the amplitude decreases more slowly in the local area near the front, i.e., like $t^{-1/3}$. Indeed, it is easy to show that for the front of the gravity wave train, the asymptotic formula previously derived must be replaced by

$$\eta = \frac{1}{\sqrt{2\pi}} \frac{N_0(0)}{D} \frac{\pi}{(c_0 t / 2D)^{1/3}} A_i \left(\frac{x - c_0 t}{D[c_0 t / 2D]^{1/3}} \right), \quad c_0 \equiv \sqrt{gD} \tag{6.6}$$

where A_i is the first Airy function that is a solution of the ordinary differential equation

$$\frac{d^2 A_i}{dx^2} - x A_i = 0 \tag{6.7}$$

so that it is oscillatory for negative values of its argument but exponentially decreasing for positive values of its argument, as shown in Fig. 6.2.

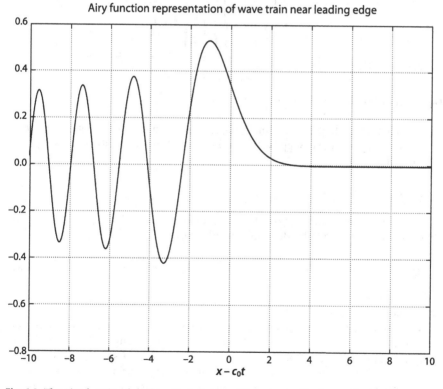

Fig. 6.2. The Airy function describing the behavior of the wave amplitude near the leading edge of the advancing wave front

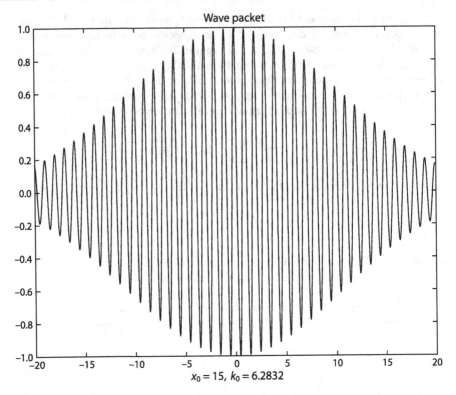

Fig. 6.3. The wave packet described by Eq. 6.8

Another example that is also illuminating occurs when the initial spatial perturbation is *nearly* a plane wave. Suppose that at time $t = 0$, the spatial distribution of η is of the form (Fig. 6.3)

$$\eta = N_0 e^{-(x/x_0)^2} \cos k_0 x \tag{6.8}$$

The wave packet is shown above. By using standard tables of integrals, it is easy to show that the Fourier amplitude of the disturbance is

$$N(k) = \frac{1}{\sqrt{2}} x_0 e^{-(k-k_0)^2 x_0^2 / 4} \tag{6.9}$$

and is shown in Fig. 6.4.

Notice that the confinement length in x is x_0, while the width of the spectrum is of order $1/x_0$. Thus, if the disturbance is broad in x, approximating a plane wave slowly modulated by the long envelope, the spectrum is very narrow in k space. This, of course, is the basic content of the quantum mechanical uncertainly principle, where k and x are the momentum and position coordinates. We do not need to get very fancy here, but it is important to note that with an $N(k)$ so sharply peaked, the formula we previously derived for the evolution of the free surface,

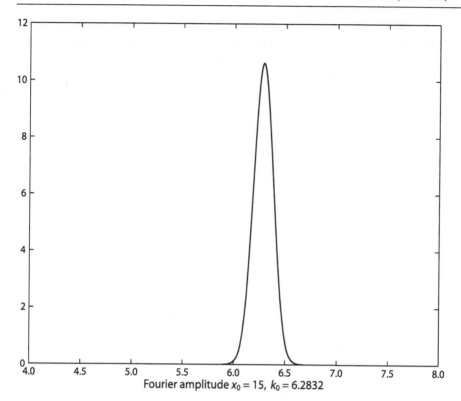

Fig. 6.4. The Fourier amplitude of the wave packet of Eq. 6.8

$$\eta(x,t) = \frac{1}{\sqrt{2\pi}} \int\limits_{-\infty}^{\infty} \frac{N_0(k)}{2} \left[e^{i\omega t} + e^{-i\omega t} \right] e^{ikx} dk \tag{6.10}$$

can be evaluated using the fact that for k distant from the spectral peak at $k = k_0$ (nothing whatever here to do with stationary phase), the integrand is essentially zero.

$$\eta = \frac{N_0}{\sqrt{2\pi}} \int\limits_{-\infty}^{\infty} e^{-(k-k_0)^2 x_0^2/4} e^{i(kx - \omega t)} dk$$

$$\approx \frac{N_0}{\sqrt{2\pi}} e^{i(k_0 x - \omega(k_0)t)} \int e^{-(k-k_0)^2 x_0^2/4} e^{i([k-k_0][x - c_g t] - i(k-k_0)^2 \omega''(k_0)/2)} dk \tag{6.11}$$

The integral is a standard one and the result is

$$\eta = \frac{x_0 N_0}{2\Delta x} e^{i(k_0 x - \omega(k_0)t)} e^{-(x - c_g(k_0)t)^2/4(\Delta x)^2} \tag{6.12}$$

where

$$(\Delta x)^2 = x_0^2/4 + i\omega''(k_0)t \tag{6.13}$$

The origin of the Gaussian packet is now centered on the position $c_g(k_0)t$, and it spreads (a little more algebra is needed to put this in real form, but the result is clear) linearly in time, yielding an amplitude that decays like the inverse of the square root of t. This is very similar to our stationary phase result for an *arbitrary* initial condition and emphasizes that the result we achieved there can be thought of as an infinite collection of packets of the type described in this idealized example.

Internal Gravity Waves

In both the atmosphere and the ocean, the fluid is density stratified, i.e., $\rho = \rho(z)$ (it is also a function of horizontal coordinates and time) so that *usually* dense fluid underlies lighter fluid. This stratification supports a new class of waves called *internal waves*. Internal waves are designated as such, because the vertical structure of the waves is oscillatory in z (contrast with the surface gravity wave) and most of the vertical displacement occurs *within* the fluid as opposed to the upper boundary, as in the gravity wave example we have just studied.

We will consider the problem in the simpler incompressible case appropriate for the ocean. The generalization to the atmosphere is straightforward if a bit more complicated (see, for example, Gill 1982 and also Lighthill 1978, for the generalization).

For an incompressible, stratified, nonrotating fluid that experiences small perturbations about a state of rest, the rest state is characterized by

$$\bar{u} = 0 , \quad \rho = \rho_0(z) , \quad p = p_0(z) \tag{7.1a--c}$$

$$\frac{\partial p_0}{\partial z} = -\rho_0 g \tag{7.1d}$$

If we examine small perturbations about a state of rest, the equations of motion, assuming that the motion is frictionless and adiabatic[1], are

$$\rho_0 \frac{\partial u}{\partial t} = -\frac{\partial p}{\partial x} \tag{7.2a}$$

$$\rho_0 \frac{\partial v}{\partial t} = -\frac{\partial p}{\partial y} \tag{7.2b}$$

$$\rho_0 \frac{\partial w}{\partial t} = -\frac{\partial p}{\partial z} - \rho g \tag{7.2c}$$

$$\frac{\partial u}{\partial x} + \frac{\partial v}{\partial y} + \frac{\partial w}{\partial z} = 0 \tag{7.2d}$$

$$\frac{\partial \rho}{\partial t} + w \frac{\partial \rho_0}{\partial z} = 0 \tag{7.2e}$$

---

[1] The student is asked to remind him(her) self what is required to make these assumptions in a consistent fashion.

where we have written the total dynamical fields for density and pressure as

$$\rho_{total} = \rho_0 + \rho, \quad \rho \ll \rho_0 \tag{7.3a}$$

$$p_{total} = p_0 + p, \quad p \ll p_0 \tag{7.3b}$$

so that all the non-subscripted variables in the above equations are perturbation quantities. Be sure to note again that the last equation is the equation for adiabatic motion, i.e., the energy equation. The condition for incompressibility is expressed by Eq. 7.2d.

A very simple special case of the above equations occurs when the horizontal velocities are *identically zero*, and when the *pressure perturbation is also zero and the vertical velocity is independent of z*. In that case, the first, second and fourth equations are trivially satisfied, and the combination of the third and fifth equation leads directly to

$$\frac{\partial^2 w}{\partial t^2} + N^2 w = 0 \tag{7.4a}$$

$$N^2 \equiv \frac{-g}{\rho_0} \frac{\partial \rho_0}{\partial z} \tag{7.4b}$$

For consistency, the quantity N must in this special case be independent of z to allow w to remain independent of z. N is called the buoyancy frequency, or sometimes the Brunt-Väisälä frequency or simply the Brunt frequency (depending on your national prejudice). Whatever it is called, the simple motion we have examined, columns of vertical motion rising or falling with no variation in the vertical direction, oscillate with the frequency N, which depends on the degree of vertical stratification. It is helpful to compare this frequency with the frequency of surface gravity waves. For deep water waves of wave number k for example (these are the relatively slow surface waves), the ratio of the surface to internal wave frequencies is

$$\frac{\omega_{int.}}{\omega_{surf.}} = \frac{N}{\sqrt{gk}} = \left(\frac{-\partial \rho_0}{k \rho_0 \partial z} \right)^{1/2} \approx \left(\frac{-\partial \rho_0}{\rho_0 \partial z} \lambda \right)^{1/2}$$

$$\approx \left(\frac{\Delta \rho_0}{\rho_0} \right)^{1/2} \tag{7.5}$$

Here we have used the fact that the vertical scale of the surface gravity wave is its wavelength λ, and that scale times the vertical derivative of the density gives an estimate of the overall change of density on that scale. Since, in the ocean, the density changes by less than 0.001 over the *total* depth, the ratio of the frequencies is such that the internal wave frequencies are always smaller than the surface wave frequencies. This makes sense, since the gravitational restoring force for surface waves depends on the difference between the density of air and water, while for the internal waves it depends on the slight difference of density between adjacent strata of fluid.

We can derive a more general equation for the internal wave field. Taking the horizontal divergence of the horizontal momentum equations yields

$$\rho_0 \frac{\partial}{\partial t} \nabla_h \cdot \vec{u}_h = -\nabla_h^2 p = -\rho_0 \frac{\partial^2 w}{\partial z \partial t} \tag{7.6}$$

with the aid of the continuity equation. In the above equation, the subscript h refers to the two-dimensional operator in the horizontal plane. z is the vertical coordinate antiparallel to gravity, and x and y are the horizontal coordinates.
    Thus,

$$\frac{\partial^2 w}{\partial z \partial t} = \nabla_h^2 p / \rho_0 \tag{7.7}$$

The time derivative of the vertical equation of motion with the aid of the adiabatic equation yields

$$\frac{\partial^2 w}{\partial t^2} + N^2 w = -\frac{1}{\rho_0} \frac{\partial^2 p}{\partial t \partial z} \tag{7.8}$$

Note that for zero pressure fluctuation, the problem reduces to the case of the oscillation at frequency N.
    Eliminating the pressure between the last two equations yields

$$\frac{\partial^2}{\partial t^2} \left[\nabla_h^2 w + \frac{1}{\rho_0} \frac{\partial}{\partial z} \left(\rho_0 \frac{\partial w}{\partial z} \right) \right] + N^2 \nabla_h^2 w = 0 \tag{7.9a}$$

$$\nabla_h^2 \equiv \frac{\partial^2}{\partial x^2} + \frac{\partial^2}{\partial y^2} \tag{7.9b}$$

Before continuing to find solutions, let's examine the last term in the square bracket on the left-hand side of the equation. This is

$$\frac{1}{\rho_0} \frac{\partial}{\partial z} \left(\rho_0 \frac{\partial w}{\partial z} \right) = \frac{1}{\rho_0} \frac{\partial \rho_0}{\partial z} \frac{\partial w}{\partial z} + \frac{\partial^2 w}{\partial z^2} \tag{7.10}$$

The ratio of the two terms on the right-hand side is

$$\frac{\partial w / \partial z \, \partial \rho_0 / \partial z}{\rho_0 \, \partial^2 w / \partial z^2} = \frac{d}{\rho_0} \frac{\partial \rho_0}{\partial z} \ll 1 \tag{7.12}$$

where d is the vertical scale of the vertical velocity w. Since that scale for internal waves is less than the total depth of the ocean (it is usually of the order of the thickness of the thermocline or less) the ratio is less than the total density change from top to bottom in the ocean, a term, again, very much less than unity. Thus, in the governing equation for density, the derivative of the background density with respect to z may be ignored, leading to the simpler governing equation:

$$\frac{\partial^2}{\partial t^2}\left[\nabla^2 w\right] + N^2\nabla_h^2 w = 0 \tag{7.13}$$

where the Laplacian following the second time derivative is now the full, three-dimensional Laplacian. Note that the structure is not spatially isotropic. The term multiplying N^2 involves only horizontal derivatives. In the presence of stratification, horizontal and vertical directions have dynamically different significance. Note also that if N is zero, we obtain, again, Laplace's equation for the vertical velocity, i.e., in the absence of stratification, the flow would be irrotational.

In fact, it is left to the student to show that the three components of the vorticity equation in this linearized example are

$$\frac{\partial}{\partial t}\left(v_x - u_y\right) = 0 \tag{7.14a}$$

$$\frac{\partial}{\partial t}\left(w_y - v_z\right) = -g\left(\frac{\rho}{\rho_0}\right)_y \tag{7.14b}$$

$$\frac{\partial}{\partial t}\left(u_z - w_x\right) = g\left(\frac{\rho}{\rho_0}\right)_x \tag{7.14c}$$

for the z-, x- and y-components of the vorticity equation (subscripts in the above equations denote partial differentiation). If the perturbation density is zero, which will occur if there were no density variation in the basic state, each component of vorticity would be zero if initially so. Thus for internal gravity waves, we can anticipate that the relative vorticity will be different from zero.

Let us try to find plane wave solutions in three dimensions, i.e., we write

$$w = w_0 \cos(\vec{K} \cdot \vec{x} - \omega t) \tag{7.15}$$

$$\vec{K} \cdot \vec{x} = kx + ly + mz$$

Inserting this trial solution in the governing partial differential equation yields as the condition for nontrivial solutions

$$\omega^2 K^2 = N^2 K_h^2 = N^2(k^2 + l^2) \tag{7.16}$$

or

$$\omega = \pm N\frac{K_h}{K} \tag{7.17}$$

Since $K^2 = k^2 + l^2 + m^2$, the frequency can be written:

$$\omega = \pm N\cos\vartheta \tag{7.18}$$

where ϑ is the angle between the wave vector K and the horizontal plane (Fig. 7.1).

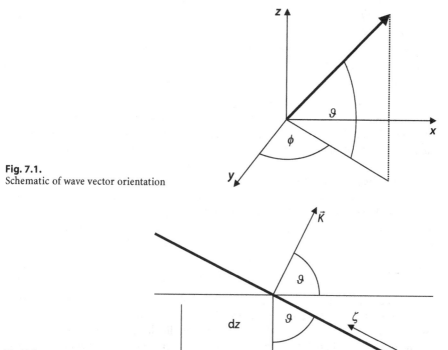

Fig. 7.1.
Schematic of wave vector orientation

Fig. 7.2.
The force diagram along
a wave crest

The frequency depends only on the orientation of the wave vector and not its magnitude. The frequency is therefore independent of the wavelength.

To get a better appreciation for the reason for this rather bizarre dispersion relation, consider a balance of forces along a line of constant phase, for example the crests of the waves, as shown in Fig. 7.2.

Let ζ be the displacement of a fluid element *along* the line of constant phase. If the wave vector is tilted to the horizontal at an angle ϑ, a displacement of an amount ζ along the phase line yields a vertical displacement $dz = \zeta \cos \vartheta$. This in turn yields a buoyancy force in the vertical direction (positive upward) of

$$F_z = -\Delta \rho g = \frac{\partial \rho_0}{\partial z} dz = \frac{\partial \rho_0}{\partial z} \zeta \cos \vartheta$$

The component of this force along the direction of the phase line is just

$$F_\zeta = \frac{\partial \rho_0}{\partial z} \zeta \cos^2 \vartheta$$

Since, by definition, there can be no variation of pressure along a phase line (nothing in the wave field varies along a line of constant phase for a plane wave), there is no pressure force along the phase line and the force balance reduces to

$$\rho_0 \frac{\partial^2 \zeta}{\partial t^2} = g \frac{\partial \rho_0}{\partial z} \zeta \quad \text{or} \tag{7.19a}$$

$$\frac{\partial^2 \zeta}{\partial t^2} + N^2 \cos^2 \vartheta \zeta = 0 \tag{7.19b}$$

which recovers our dispersion relation for frequency of a harmonic oscillation. Notice that when ϑ is 0, we recover the first simple case in which the frequency of oscillation is exactly N. To understand the reason for *that*, note that for a plane wave, such that all fields are of plane wave type

$$(u,v,w) = (u_0,v_0,w_0)e^{i(kx+ly+mz-\omega t)} \tag{7.20}$$

the continuity equation imposes the condition

$$ku_0 + lv_0 + mw_0 = 0 \quad \text{or} \tag{7.21a}$$

$$\vec{K} \cdot \vec{u} = 0 \tag{7.21b}$$

Thus, the fluid velocity in the three-dimensional plane wave is *perpendicular* to the wave vector. The fluid velocity is *along* the crests of the waves, i.e., for internal waves, the wave motion is *transverse*; that is, it is perpendicular to the direction of phase propagation. Thus when the wave vector is horizontal, the motion of fluid elements in the wave is purely vertical and with no variation of phase in z ($m = 0$) the vertical motion will be independent of z. Those were the conditions of our introductory example, and we see here that this is obtained when the wave vector is horizontal. It also yields the maximum frequency for internal waves, i.e. $\omega_{max} = N$.

Note too that the frequency is a constant on a *cone* in three-dimensional wave number space (Fig. 7.3) where the elements of the cone make an angle ϑ to the horizontal. The frequency increases as the cone opens up, i.e., when the elements of the cone are closer to the horizontal plane. This has important consequences for the direction of the group velocity, since the frequency is a function only of ϑ.

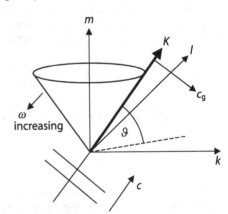

Fig. 7.3.
The cone of constant frequency and the direction of the group velocity

Group Velocity for Internal Waves

By definition, the three-dimensional group velocity is

$$\vec{c}_g = \hat{i}\frac{\partial \omega}{\partial k} + \hat{j}\frac{\partial \omega}{\partial l} + \hat{k}\frac{\partial \omega}{\partial m} \tag{7.22}$$

where $\hat{i}, \hat{j}, \hat{k}$ are the three unit vectors along the x-, y- and z-axes, respectively. Let ϕ be the angle in the x-y-plane between the horizontal projection of the wave vector and the x-axis (Fig. 7.4).
Then a simple calculation using

$$\omega^2 = N^2 \frac{k^2 + l^2}{k^2 + l^2 + m^2} \tag{7.23}$$

yields

$$\frac{\partial \omega}{\partial k} = \frac{N}{K}\frac{m^2}{K^2}\frac{k}{K_h} = \frac{N}{K}\sin\vartheta\{\sin\vartheta\cos\phi\} \tag{7.24a}$$

$$\frac{\partial \omega}{\partial l} = \frac{N}{K}\frac{m^2}{K^2}\frac{l}{K_h} = \frac{N}{K}\sin\vartheta\{\sin\vartheta\sin\phi\} \tag{7.24b}$$

$$\frac{\partial \omega}{\partial m} = -N\frac{K_h m}{K^3} = -\frac{N}{K}\cos\vartheta\sin\vartheta \tag{7.24c}$$

In particular, note that

$$\frac{\omega}{m}\frac{\partial \omega}{\partial m} = -\frac{N^2}{K^2}\cos^2\vartheta \tag{7.25}$$

so that the vertical phase velocity is always opposite to the vertical group velocity. Waves that appear to be propagating their phase upwards will be propagating their energy downwards, and vice versa. This is evident from examining the dispersion cone in three dimensions, keeping in mind that the frequency increases in a direction perpendicular to the elements of the cone as shown in Fig. 7.3.

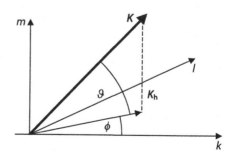

Fig. 7.4.
The orientation of the wave vector

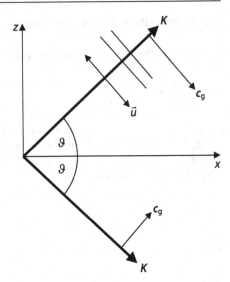

Fig. 7.5.
The orientation of the group velocity with
respect to the wave vector

Finally, note that

$$\vec{K}\cdot\vec{c}_g = kc_{gx} + lc_{gy} + mc_{gz} = \frac{N^2 m^2}{\omega K^4}(k^2 + l^2 - K_h^2) = 0 \tag{7.26}$$

so that the *group velocity is perpendicular to the wave vector and therefore in
the direction of the fluid velocity*. Energy travels *along the crests and troughs* and
not perpendicular to them. For surface gravity waves, we had to get used to the fact that
the group velocity was not equal in magnitude to the phase speed. Now, for internal
gravity waves, we have to adjust to the remarkable fact that the group velocity is not
even in the same direction as the propagation of phase but at right angles to it (Fig. 7.5).

Internal Waves, Group Velocity and Reflection

The rather unusual dispersion relation and the nonintuitive relation between group velocity and the wave vector lead to some very unusual physical consequences.

Figure 8.1 is from Lighthill's book (1978) taken from a paper by Mowbray and Rarity (1967). It shows the result of an experiment in which a small disk is oscillated in a stratified fluid with a constant N at a constant frequency, ω. We know that in such a case the wave vectors will be aligned in a direction such that $\cos \vartheta = \pm \omega / N$. There are four such angles. The disturbance is limited to narrow bands leading away from the oscillating disk. Since the energy must be moving away from the disk, it is not too hard to see that starting with the band in the upper right hand quadrant, the direction of the band must correspond to the direction of the outgoing group velocity moving upward and to the right. Since this must be perpendicular to the wave number, and since the vertical group velocity and phase speeds must be oppositely directed, it follows that the dark bands in the figure are actually the *crests* of the internal waves, which form a cross intersecting the little disk. A movie of the experiment would show those crests moving rightward and downward in the upper right band corresponding to energy moving upward and to the right. The situation is sketched schematically at the right side. The student is invited to complete the picture for the other four quadrants. One has to admit that the physics here seems very strange. But you'll get used to it.

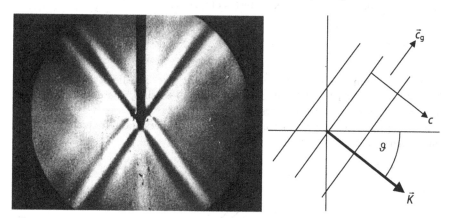

Fig. 8.1. A photograph showing the lines of constant phase produced by a small disk oscillating with constant frequency. Below a diagram is shown, indicating the lines of constant phase and the direction of the group velocity of the radiated waves (from Lighthill 1978)

Note that since the fluid element velocities are at right angles to the wave vector, the pressure work term $p\bar{u}$ is perpendicular to the wave vector. Since we expect the energy flux to be given by the pressure work term, this explains the perpendicular orientation of K and $\bar{c}_g$ (Can you derive the energy equation from the momentum and thermodynamic equations we used at the beginning?).

Something even stranger appears to happen if we ask about the reflection of internal gravity waves from a solid boundary. Let us suppose we have a lower boundary sloping upward to the right at an angle β. We will suppose the incident wave and reflected wave are in the plane of the slope. It is easy to consider the generalization in Fig. 8.2.

Suppose the incident wave has the representation

$$w = W_I e^{i(k_I x + m_I z - \omega_I t)} \tag{8.1}$$

where the I subscripts refer to the *incident wave field*.

For this two dimensional problem (no y wave number), the continuity equation is simply

$$\frac{\partial u}{\partial x} + \frac{\partial w}{\partial z} = 0 \tag{8.2}$$

so that a stream function can be used, where

$$u = -\frac{\partial \psi}{\partial z} \tag{8.3a}$$

$$w = \frac{\partial \psi}{\partial x} \tag{8.3b}$$

and the incident wave, represented in terms of its stream function is

$$\psi_I = \frac{W_I}{ik_I} e^{i(k_I x + m_I z - \omega_I t)} = \Psi_I e^{i(k_I x + m_I z - \omega_I t)} \tag{8.4}$$

The solid boundary at which the reflection takes place satisfies $z = x \tan\beta$ so that the unit vector parallel to the boundary is $\hat{i}_B = \hat{x} \cos\beta + \hat{z} \sin\beta$, where $\hat{x}$ and $\hat{z}$ are unit vectors in the x- and z-direction, respectively.

Now the reflected wave will have x and z wave numbers and a frequency that are not known a priori. How are they determined?

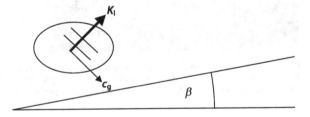

Fig. 8.2.
A wave packet with wave number K_I is incident on a sloping surface

We can write the reflected wave, generally, as

$$\psi_r = \Psi_r e^{i(k_r x + m_r z - \omega_r t)} \tag{8.5}$$

The total wave field during the reflection process is the sum of the two waves,

$$\psi = \psi_I + \psi_r \tag{8.6}$$

On the boundary where $z = x \tan\beta$, the total stream function must be a constant. Without loss of generality, let that constant be zero. Thus, on $z = x \tan\beta$ we have

$$0 = \Psi_I e^{i[(k_I + m_I \tan\beta)x - \omega_I t]} + \Psi_r e^{i[(k_r + m_r \tan\beta)x - \omega_r t]} \tag{8.7}$$

This must be true for all t and for all x *along the boundary*. Clearly a single relation between the amplitudes of the incoming and reflected waves will be unable to satisfy Eq. 8.7, *unless*

$$\omega_I = \omega_r \tag{8.8a}$$

$$k_I + m_I \tan\beta = k_r + m_r \tan\beta \quad \text{or} \tag{8.8b}$$

$$\vec{K}_I \cdot \hat{i}_B = \vec{K}_r \cdot \hat{i}_B \tag{8.8c}$$

Thus, *the frequency and the component of the wave vector parallel to the boundary are both conserved under reflection.* This is a general result for plane waves. What is special for internal gravity waves is that the conservation of frequency implies that since $\omega = N \cos\vartheta$, the angle of the wave vector with the horizontal must be preserved under reflection, regardless of the orientation with the boundary. For more familiar problems where the reflection is specular, the wave vector component perpendicular to the boundary is preserved. This is not the case here. Thus, under reflection both the component of wave vector *along* the boundary and the horizontal component of the wave vector must be preserved.

We can use a geometrical construction to see how this occurs (see Fig. 8.3).

In the construction, the reflected wave vector is determined by three considerations:

1. The component along the slope must be the same for both incident and reflected wave;
2. The angle of the reflected wave vector to the horizontal must have the same magnitude as for the incident wave so that the cosine of the angle (frequency) is preserved under reflection;
3. The direction of the reflected wave vector must be such that the associated group velocity is directed *away* from the slope.

We note that in this example, the magnitude of the reflected wave vector is much greater than that of the incident wave. Therefore, the wavelength of the wave is *not* preserved under reflection; indeed, the *wavelength shortens as a consequence of the reflection process.*

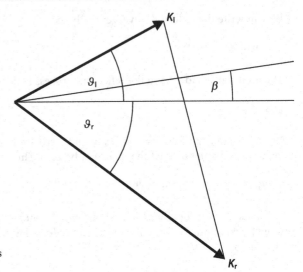

Fig. 8.3.
A sketch showing the wave
numbers of the incident and
reflected internal gravity waves
from a sloping surface

To determine the result analytically, let the magnitude of the incident wave vector
be K_I. Then the component of the incident wave vector along the slope is $K_I \cos(\vartheta - \beta)$,
while that of the reflected wave along the slope is $K_r \cos(\vartheta + \beta)$. Note that we have used
the fact that ϑ is preserved under reflection. Since these two terms must be equal to
satisfy the condition of no flow through the solid surface of the slope,

$$K_I \cos(\vartheta - \beta) = K_r \cos(\vartheta + \beta) \tag{8.9}$$

Now define

$$\alpha = \pi / 2 - \vartheta$$

Here α is the angle with respect to the horizontal of the group velocity of the incident packet and also the angle with respect to the horizontal of the reflected packet
(see Fig. 8.4). Note too that $\omega = N \sin \alpha$.

If the definition of α is used in the equation for the equality of the along-slope components of the wave vectors, we obtain

$$\frac{K_r}{K_I} = \frac{\sin(\alpha + \beta)}{\sin(\alpha - \beta)} \tag{8.10}$$

Note that as the angle of the incident group velocity approaches the angle of the
slope, the magnitude of the reflected wave number becomes infinite. Thus, as the frequency is lowered, α gets smaller; when it coincides with β, K_r becomes infinite. We
would anticipate that such short scales would be affected by friction and are likely to
be efficiently dissipated. So, β is a critical angle for the incident packet.

There is also a problem when α is less than β according to the above formula; since
neither of the wave number magnitudes can be negative, the left-hand side is always
positive, but the right-hand side becomes negative when $\alpha < \beta$. Clearly, the situation

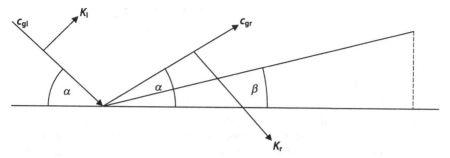

Fig. 8.4. A sketch showing the relation of the incident and reflected wave vectors

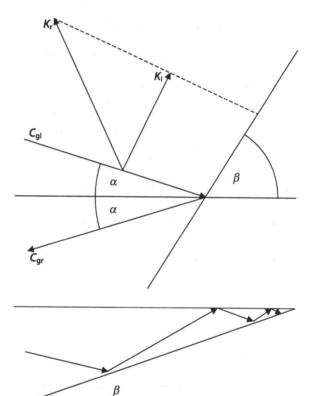

Fig. 8.5.
The reflection process when
the slope is steep ($\alpha < \beta$)

Fig. 8.6.
The reflection in a shallow
wedge

must be reconsidered in that case. Figure 8.5 shows the geometry of the reflection process, then (see also Fig. 8.6). We see that when $\alpha > \beta$, as in the previous case, the reflection is *forward* along the slope. Now, when $\alpha < \beta$, the reflection must be *backward* (and
since α is preserved, forward reflection would put the wave packet *inside* the slope,
which is an impossibility). The back reflection leads to the relation (try it)

$$\frac{K_r}{K_I} = \frac{\sin(\alpha + \beta)}{\sin(\beta - \alpha)} \quad \text{a result one might have guessed.}$$

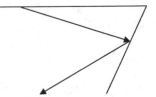

Fig. 8.7.
The reflection in a large, open
wedge

Since the reflection from a horizontal surface will be specular if the bottom slope forms a wedge-shaped region with an upper horizontal surface, the reflection process can lead to further surprises. If, for example, a wave packet enters the wedge with a frequency such that $\alpha > \beta$, the reflection from both the bottom slope and the top surface will be forward. The wave packet will bounce back and forth, advancing towards the apex of the wedge, becoming shorter at every bottom reflection, and finally dissipating in the apex of the wedge (Fig. 8.7).

If, on the other hand, the bottom slope is strong enough so that $\beta > \alpha$ (in the limit it could be a vertical wall), the reflection from the bottom will be backward, and the wave will leave the region of the wedge (Fig. 8.7).

Recall that all of these bizarre properties are due entirely to the fundamental physics of the wave that determines its frequency only in terms of the angle the wave vector makes to the horizontal, and since the frequency must be preserved under reflection, this places a terrific constraint on the reflection kinematics.

Up to now we have dealt with fluids in which the buoyancy frequency has been independent of z. In the ocean, N is certainly a function of z. It is large in the thermocline and small in both the mixed layer and in the abyss. Before dealing with that variation, it is useful to discuss the equation for the energy in the wave field.

Multiplying each momentum equation by the velocity component in that direction, we obtain

$$\frac{\partial KE}{\partial t} = -u\frac{\partial p}{\partial x} - v\frac{\partial p}{\partial y} - w\frac{\partial p}{\partial z} - w\rho g \tag{8.11}$$

$$= \nabla \cdot p\vec{u} - w\rho g$$

where

$$KE = \frac{\rho_0}{2}\left[u^2 + v^2 + w^2\right] \tag{8.12}$$

The step between the first and second equations uses the condition of incompressibility, i.e., the divergence of the velocity vanishes. The last term on the right-hand side of the equation for the kinetic energy is the transformation of potential to kinetic energy. If heavy fluid sinks ($\rho > 0$, $w < 0$) or light fluid rises ($\rho < 0$, $w > 0$) (where we recall that ρ is the density *perturbation*), then the kinetic energy will increase by the conversion of gravitational potential energy. This last term can be written in conservation form using the adiabatic equation, since from that equation it follows that

$$w\rho g = \frac{\rho}{\rho_0}\frac{g^2}{N^2}\frac{\partial \rho}{\partial t} = \frac{g^2}{2\rho_0 N^2}\frac{\partial \rho^2}{\partial t} \tag{8.13}$$

Thus,

$$\frac{\partial}{\partial t}\left[KE + g^2\rho^2/2\rho_0 N^2\right] + \nabla \cdot p\vec{u} = 0 \tag{8.14}$$

It is tempting to consider the second term in the square bracket as potential energy. To see that in fact it is, it is useful to use the relationship between the Lagrangian vertical displacement ζ and the vertical velocity. For small displacements,

$$w = \frac{\partial \zeta}{\partial t} \tag{8.15}$$

(Large displacements would require the total derivative in the above equation).
   If this is used in the adiabatic equation, we obtain

$$\rho = \rho_0 \frac{N^2}{g}\zeta \tag{8.16}$$

by a simple integration. We can therefore think of ζ as the vertical displacement of each isopycnal surface, such that the perturbed fluid element remains on its original density surface. In turn, we can now write the energy equation as

$$\frac{\partial}{\partial t}\left[KE + \frac{\rho_0 N^2 \zeta^2}{2}\right] + \nabla \cdot p\vec{u} = 0 \tag{8.17}$$

so that the second term in the square bracket has exactly the same form as the potential energy of an extended spring in which the spring constant per unit mass measuring the restoring force is the buoyancy frequency squared, i.e., N^2.
   It will be left for the student to show that for a plane wave, there is equipartition between kinetic and potential energy and that the energy flux vector

$$p\vec{u} = \vec{c}_g E \tag{8.18}$$

where E is the sum of the kinetic and potential energy.
   Note that for a plane wave in two dimensions, we can always align our coordinate system for a single plane wave to align the wave vector in the x-z-plane.
   Suppose the plane wave has the form

$$w = w_0 \cos\Theta \tag{8.19a}$$

$$\Theta = kx + mz - \omega t \tag{8.19b}$$

Then from the continuity equation,

$$u = w_0 \frac{m}{k}\cos\Theta \tag{8.20}$$

(note that this satisfies the condition that the fluid velocity be perpendicular to the wave vector). From the relation between w and the vertical displacement ζ, (or from the adiabatic equation),

$$\zeta = -\frac{w_0}{\omega}\sin\Theta \tag{8.21}$$

Thus, the kinetic and potential energies averaged over a wave period are

$$\langle E \rangle = \frac{\rho_0}{4}w_0^2\left\{1+\frac{m^2}{k^2}+\frac{N^2}{\omega^2}\right\} \tag{8.22a}$$

$$\langle E \rangle = \frac{\rho_0}{2}w_0^2\left\{\frac{k^2+m^2}{k^2}\right\} \tag{8.22b}$$

from which it follows that the horizontal and vertical components of the energy flux are

$$c_{gx}\langle E \rangle = \frac{\rho_0 w_0^2}{2}\left(\frac{\omega}{k}\right)\frac{m^2}{k^2} \tag{8.23a}$$

$$c_{gz}\langle E \rangle = -\frac{\rho_0 w_0^2}{2}\left(\frac{\omega}{m}\right)\frac{m^2}{k^2} \tag{8.23b}$$

Note again that the direction of the vertical energy flux is opposite to that of the vertical phase speed ω/k. Indeed, the energy flux is perpendicular to the wave vector as can be immediately verified.

WKB Theory for Internal Gravity Waves

The buoyancy frequency is never really constant. Indeed, in the ocean there is a significant variation of N from top to bottom. Figure 9.1 (next page) from the *Levitus Atlas* (1982) shows the distribution of N of the zonally averaged global ocean.

By assuming that N is constant in our calculations to this point, we have been saying effectively that over the vertical distance $\lambda_z = 2\pi/m$, N^2 changes only slightly. Alternatively, we can state equivalently that N is a slowly varying function with respect to the phase of the wave over which

$$\delta N^2 = \frac{\partial N^2}{\partial z}\lambda_z \ll N^2 \tag{9.1}$$

The major change in N occurs in the vertical; that is, it is a stronger function of z than of the horizontal coordinates. We already know from our earlier work on waves in slowly varying media that

$$\frac{\partial \vec{K}}{\partial t} + \vec{c}_g \cdot \nabla \vec{K} = -\nabla \Omega \tag{9.2}$$

where Ω is the local dispersion relation. If we consider N to be a function only of z, this equation implies that the vertical component m will be the only component of the wave vector that will alter as the wave traverses a region of variable N. Furthermore, if we assume that N is independent of time, the dual equation for the frequency

$$\frac{\partial \omega}{\partial t} + \vec{c}_g \cdot \nabla \omega = \frac{\partial \Omega}{\partial t} \tag{9.3}$$

shows that the frequency will be independent of time for an observer moving with the group velocity. Since the wave packet itself will move with the group velocity, this means that (k, l, ω) will be constant. It remains to be determined how m changes and how the amplitude will vary in a region of varying N.

The governing equation is again

$$\frac{\partial^2}{\partial t^2}\nabla^2 w + N^2(z)\nabla_h^2 w = 0 \tag{9.4}$$

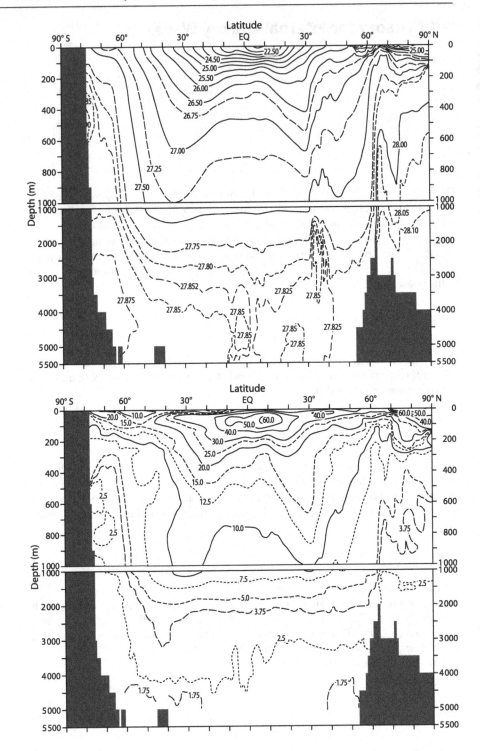

Let's try to find a solution in the form

$$w = A(z)e^{i(kx+ly-\omega t+\theta(z))} \qquad (9.5)$$

and we will assume that the vertical variation of the phase is much larger that the vertical variation of A; that is, we will assume that N is varying slowly enough in z so that locally our solution will look like a plane wave.
We define

$$m(z) = \frac{\partial \theta}{\partial z} \qquad (9.6)$$

(Note that for a pure plane wave, θ would be simply mz).
Inserting the hypothesized solution in the governing equation yields (z subscripts denote differentiation)

$$-\omega^2 \left[-(k^2+l^2)A + A_{zz} - \theta_z^2 A \right] - N^2(k^2+l^2)A - i\omega^2(2\theta_z A_z + \theta_{zz}A) = 0 \quad \text{or} \qquad (9.7)$$

$$A_{zz} + A\left(\frac{(N^2-\omega^2)K_h^2}{\omega^2} - \theta_z^2 \right) - 2i\omega^2\theta_z^{1/2}\frac{\partial}{\partial z}\left(\theta_z^{1/2}A \right) = 0 \qquad (9.8)$$

We have assumed that θ_z is order one while $A_{zz}/A \ll 1$ and that the variation of θ_z and A are also small (the local plane wave approximation. This implies that the dominant term in the equation is the curved bracket in the second term. This yields an expression for m or equivalently,

$$\theta_z^2 = m^2 = \frac{N^2-\omega^2}{\omega^2}K_h^2 \quad \text{or} \qquad (9.9a)$$

$$m(z) = \frac{\partial \theta}{\partial z} = \left[\frac{N^2(z)-\omega^2}{\omega^2} \right]^{1/2} K_h \qquad (9.9b)$$

or, equivalently, one may think of this relation as the necessary condition that the frequency both satisfy the plane wave dispersion relation

$$\omega^2 = \frac{N^2(z)K_h^2}{m^2+K_h^2} \qquad (9.10)$$

while at the same time be independent of z. This yields for the vertical phase factor

$$\theta = \int_{z_0}^{z} K_h\sqrt{\frac{N^2(z')-\omega^2}{\omega^2}}\,dz' \qquad (9.11)$$

Fig. 9.1. *Upper panel: Annual mean global potential density distribution in depth and latitude for the world ocean. Lower panel: Annual mean of the buoyancy frequency as a function of latitude and depth (reworked after Levitus 1982)*

With the differential equation in the form

$$\frac{d^2W}{dz^2} + \frac{N^2 - \omega^2}{\omega^2} K_h^2 W = 0 \tag{9.12}$$

we might naively have expected the vertical structure for slowly varying N to look like

$$e^{i\left(K_h\sqrt{\frac{N^2 - \omega^2}{\omega^2}}\right)z}$$

but instead it is the integral that enters the phase so that the vertical component of the wave number vector is given by its local plane wave value.

The imaginary part of the equation for A (or equivalently, the next order term is the slow variation with z) yields the constraint

$$\frac{\partial}{\partial z}\left[Am^{1/2}\right] = 0 \tag{9.13}$$

or

$$A(z) = \frac{A(z_0)}{(m/m_0)^{1/2}} \tag{9.14}$$

so that z_0 and m_0 are evaluated at some arbitrary constant value of depth. As m gets larger, i.e., in a region of larger N, the amplitude diminishes. This is easy to understand physically. As the wave propagates vertically, the flux of energy must remain the same at each z to avoid the pile up of energy and the local increase of amplitude with time. We saw in the last lecture that the vertical energy flux could be written

$$c_{gz}\langle E \rangle = -\frac{\rho_0 A^2}{2}\frac{m\omega}{k^2}$$

using A instead of w_0 for the amplitude. To keep the energy flux independent of z, and since both frequency and horizontal wave number are independent of z, it follows that A must go inversely with $m^{1/2}$, which is the result we have already achieved. Thus that behavior is simply a consequence of energy conservation.

If the wave propagates to an elevation where the frequency is greater than the local value of N, the vertical wave number becomes purely imaginary and the disturbance exponentially decays beyond that location. We can say then that the wave will be trapped between regions in z where N matches the frequency of the wave.

To find the path of the wave packet in the vicinity of such a region, we can use the *ray equations*, which for two dimensions are

$$\frac{dz}{dt} = c_{gz} = -N\frac{km}{K^3} \tag{9.15a}$$

$$\frac{dx}{dt} = c_{gx} = N\frac{m^2}{K^3} \tag{9.15b}$$

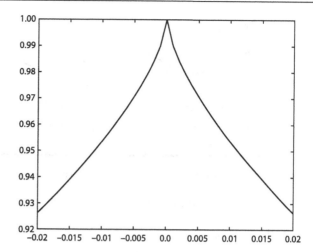

Fig. 9.2.
The ray path near a turning
point where the frequency ω
equals N

The path of the ray in the x-z-plane is given by

$$\frac{dz}{dx}=\frac{c_{gz}}{c_{gx}}=-k/m=-\frac{\omega}{\sqrt{N^2-\omega^2}} \tag{9.16}$$

Consider the region near the point z^* where $N(z^*)=\omega$. Expanding N^2 around z^* yields

$$N^2(z)=N^2(z^*)-(z^*-z)(N^2)'+\dots \tag{9.17a}$$

$$\frac{dz}{dx}=\frac{\omega}{\sqrt{(-dN^2/dz)_{z^*}(z^*-z)}} \quad\text{or} \tag{9.17b}$$

$$z^*-z=\frac{\omega}{\sqrt{-(dN^2/dz)_{z^*}}}(x^*-x)^{2/3} \tag{9.18}$$

The ray path has a cusp at the turning point (x^*, z^*) as shown in Fig. 9.2.

Normal Modes (Free Oscillations)

Consider a fluid bounded below by a flat bottom at $z=-D$ and with a free surface whose
rest position is $z=0$ (Fig. 9.3). Again, the fluid is incompressible and stratified. This
situation is a combination of the two problems previously studied. There should be
the possibility of surface gravity waves as well as internal waves due to the stratification. The issue here is how they relate to each other and in addition, what the nature
of the internal waves in this bounded domain is.

Again, the governing equation is

$$\frac{\partial^2}{\partial t^2}\nabla^2 w+N^2(z)\nabla_h^2 w=0 \tag{9.19}$$

for unforced motions.

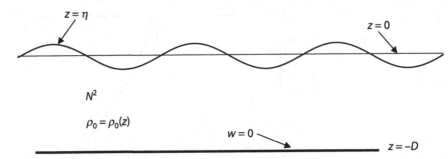

Fig. 9.3. The definition figure for determining the normal modes of a stratified fluid with a free surface

The boundary conditions are

1. At the bottom,

$$w = 0, \quad z = -D \tag{9.20}$$

2. At the free surface, we have both the kinematic condition

$$w = \frac{\partial \eta}{\partial t} \tag{9.21a}$$

where η is the free surface elevation, and

$$p(x, y, z = \eta) = 0 \tag{9.21b}$$

Since η is supposed to be small (linear, small amplitude motions),

$$p(x, y, z = \eta) = p(x, y, 0) + \frac{\partial p}{\partial z}\eta + \ldots \tag{9.22}$$

We only want to keep terms that are linear in the amplitude of the motion on the right-hand side of the above equation, since we are doing a consistent linearization of the dynamics. Since the linearized form of the vertical momentum equation is

$$\frac{\partial p}{\partial z} = -(\rho_0 + \rho)g - \rho_0 \frac{\partial w}{\partial t} \tag{9.23}$$

each term in the above equation is of the order of the amplitude of the motion and so would yield a quadratic term when multiplied by η in the expansion of the boundary condition, except the first term on the right-hand side of the equation for the vertical pressure gradient, which yields $\partial p / \partial z$ in the absence of motion. Thus, considering only terms that are *linear* in the perturbation amplitude, the pressure condition on the free surface is

$$p(x, y, z = \eta) = 0 = p(x, y, 0) - \rho_0 g \eta \tag{9.24}$$

It is important to realize that we are *not* assuming the motion is hydrostatic; all we have used is the small amplitude approximation. We did not have to work very hard in the surface gravity wave problem, because we had the Bernoulli equation at our disposal to use at the surface. Since the fluid is stratified, the motion is no longer irrotational and the velocity potential and resulting Bernoulli equation no longer exist.

Thus at the upper boundary,

$$p(x,y,0) = \rho_0 g \eta, \quad z = 0 \tag{9.25a}$$

$$\frac{\partial \eta}{\partial t} = w \tag{9.25b}$$

Eliminating the free surface elevation between the two conditions,

$$\frac{\partial}{\partial t} p(x,y,0) = \rho_0 g w, \quad z = 0 \tag{9.26}$$

We can operate on the above equation with the *horizontal Laplacian*

$$\nabla_h^2 \frac{\partial p}{\partial t} = \rho_0 g \nabla_h^2 w \tag{9.27}$$

and use the previously derived relation (from the divergence of the horizontal momentum equations and the continuity equation)

$$\nabla_h^2 p = \rho_0 \frac{\partial^2 w}{\partial z \partial t} \tag{9.28}$$

to obtain for the upper boundary condition in terms of w:

$$\frac{\partial^2}{\partial t^2} \frac{\partial w}{\partial z} - g \nabla_h^2 w = 0, \quad z = 0 \tag{9.29}$$

We will particularly be interested in the oceanographically relevant case where the parameter $DN^2/g \ll 1$. This parameter can be interpreted in several ways. First of all, it gives a measure, as we have seen before, of the total density difference over the depth of the fluid divided by the mean density. This is very small for the ocean. Second, using our previous results, it can be seen as the ratio of the (square) of the maximum internal gravity wave frequency to the surface wave frequency (squared) for a wave whose wavelength is of the order of the depth of the fluid. We are interested, as noted, in the case when this ratio is small, i.e., when *the surface waves have higher frequency and phase speeds than the internal waves*. This helps separate the two wave types that are described by the same set of equations given above. There is a hint then that in the case when $DN^2/g \ll 1$, approximations to the governing equation will be in order if we want to concentrate on one or the other of the waves.

Since the fluid is unbounded in the x and y-directions in this simple model, we can look for plane wave solutions in the horizontal direction. Orienting the x-axis to be in the direction of the horizontal wave number, this implies that we can look for solutions of the form

$$w = W(z)e^{i(kx-\omega t)} \tag{9.30}$$

and insertion in the equation for w and its boundary conditions yields

$$\frac{d^2W}{dz^2} + k^2 \left[\frac{N^2}{\omega^2} - 1 \right] W = 0 \tag{9.31a}$$

$$W = 0, \quad z = -D \tag{9.31b}$$

$$\omega^2 \frac{dW}{dz} - gk^2 W = 0, \quad z = 0 \tag{9.31c}$$

Consider the case where N is constant and $N^2 > \omega^2$. The solution of the W equation will be oscillatory in z and the solution that satisfies the boundary condition at $z = -D$ will be

$$W = A \sin m(z + D) \tag{9.32a}$$

$$m^2 = k^2 \left[\frac{N^2}{\omega^2} - 1 \right] \tag{9.32b}$$

Note that the latter definition implies that were m known, the corresponding frequency would be

$$\omega = \pm \frac{Nk}{\sqrt{k^2 + m^2}} \tag{9.33}$$

which is a familiar result from our work on plane internal gravity waves. We can expect the above *eigenvalue problem* to yield quantized values of m so that the equation for the frequency in terms of k and m will be as in the plane wave case except that *m will no longer be a continuous variable but quantized.*

The upper boundary condition yields the *eigenvalue relation*:

$$\omega^2 m \cos mD = gk^2 \sin mD \tag{9.34}$$

or using the relationship between frequency and wave number written above,

$$\frac{N^2 m}{k^2 + m^2} = g \tan mD \tag{9.35}$$

It is useful to write the above condition in terms of *non-dimensional* wave numbers.

Let

$$m' = mD$$
$$k' = kD \tag{9.36}$$

Then the eigenvalue relation becomes

$$\left[\frac{N^2 D}{g}\right] \frac{m'}{k'^2 + m'^2} = \tan m' \tag{9.37}$$

The roots of this can be found numerically. A graph of each side of the equation is helpful in understanding the results (Fig. 9.4).

Figure 9.4 shows both the left- and right-hand sides of the dispersion relation for the case where the parameter $N^2 D / g$ is artificially large (0.1) and $kD = 1$. Still, the roots of the relation corresponding to the intersections of the two curves are very close to an integral multiple of π. In the above case, the first two are at $mD = 3.1562$ and 6.2947. For smaller values of the parameter $N^2 D / g$, it is not possible to distinguish the curve of the left-hand side of the equation from zero, in which case we can see either graphically or analytically from the equation itself that in the limit $N^2 D / g \longrightarrow 0$, the roots approach

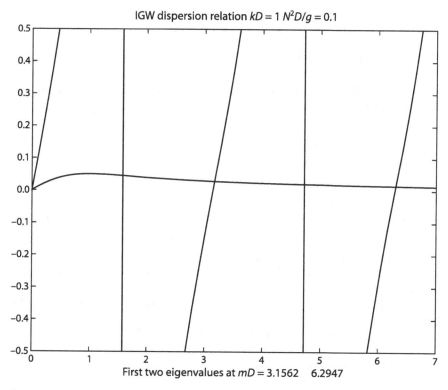

Fig. 9.4. The dispersion relation 9.37

$$mD = j\pi, \quad j = 1,2,3\ldots \tag{9.38a}$$

$$\omega = \omega_j = \pm \frac{kN}{k^2 + j^2\pi^2/D^2} \tag{9.38b}$$

In this limit, $W(z)$ is very nearly

$$W(z) \approx A\sin\frac{j\pi}{D}(z+D) = A(-1)^j \sin\frac{j\pi z}{D} \tag{9.39}$$

so that *w vanishes on both the lower and upper surfaces.* For these internal gravity wave modes for which $\omega < N$, the free surface dynamically acts as if it were rigid and the eigensolutions are the same *as if the upper surface were simply a rigid lid on which w = 0.*

Now let's examine if there are solutions of the problem for $\omega > N$. If that is the case, we can still use the same solutions and dispersion relation but we must realize that m will now be purely imaginary. It might be clearer to just go back and rewrite the solution in terms of real variables. So for $\omega > N$, we write

$$q^2 = k^2\left[1 - \frac{N^2}{\omega^2}\right] \tag{9.40}$$

in terms of which the solution satisfying the lower boundary condition is

$$w = A\sinh q(z+D) \tag{9.41}$$

which should look familiar from our surface wave studies. The upper boundary condition now yields

$$\left(\frac{N^2D}{g}\right)\frac{q'^2}{k'^2 - q'^2} = \tanh q' \tag{9.42a}$$

$$q' = qD \tag{9.42b}$$

$$\omega^2 = \frac{N^2k^2}{k^2 - q^2} \tag{9.42c}$$

When the parameter N^2D/g is small, the left-hand side will be small except in the vicinity of $q' = k'$, which yields the only eigenvalue for which $\omega > N$ (see Fig. 9.5).

This is the graph of the two sides of the eigenvalue relation when $N < \omega$. There is a single root for qD which in the case when $N^2D/g = 0.01$ is equal to $0.9884kD$, q is very nearly k.

This yields a frequency using the upper boundary condition:

$$\omega^2 q \cosh qD = gk^2 \sinh qD \tag{9.43}$$

or since q is very nearly k,

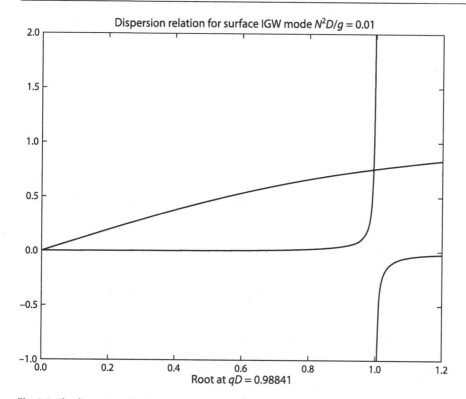

Fig. 9.5. The dispersion relation for the external mode

$$\omega \approx \pm\sqrt{gk\tanh kD} \tag{9.44}$$

and the eigenfunction is,

$$W(z) = \sinh k(z + D) \tag{9.45}$$

Both the eigenfunction and the eigenvalue in this limit are precisely the values obtained for the surface gravity wave problem for a homogeneous fluid.

Thus, the full spectrum of oscillatory modes splits into two (unequal) parts. There is first of all the free surface mode, which, when $N^2D/g \ll 1$, does not even notice the stratification. This is because the depth of penetration is of the order of the wavelength, and for small values of N^2D/g, the fluid motion in the wave, *maximum* at the surface and exponentially decreasing into the fluid, does not encounter the density variation. The second class of solutions whose frequencies are all less than N consist of internal gravity waves whose vertical motion at the upper surface is negligible. Compared to w within the fluid, the vertical velocity at the free surface is negligible and the frequencies of the modes and their structures are approximately those for a fluid contained within two horizontal rigid surfaces. It is left to the student to show that for eigenfunctions of the same amplitude, the free surface displacement of the internal gravity wave compared to that of the free surface wave is small, and this smallness is of the ratio of the respective frequencies.

For the internal gravity wave part of the spectrum using the rigid lid approximation,

$$w = W(z)e^{i(kx-\omega t)} \tag{9.46}$$

where W satisfies

$$\frac{d^2 W}{dz^2} + k^2 \left[\frac{N^2}{\omega^2} - 1 \right] W = 0 \tag{9.47a}$$

$$W = 0, \quad z = -D, 0 \tag{9.47b}$$

For N constant, the eigensolutions are the sine functions $\sin(j\pi z / D)$ and for the jth mode,

$$\omega_j / N = kD / \sqrt{k^2 D^2 + j^2 \pi^2} \quad j = 1, 2, 3 \ldots \tag{9.48}$$

Note that for large k, the frequencies of *all the modes* approach N (the student should think about the dispersion relation for plane internal gravity wave modes to understand why this is so)(Fig. 9.6).

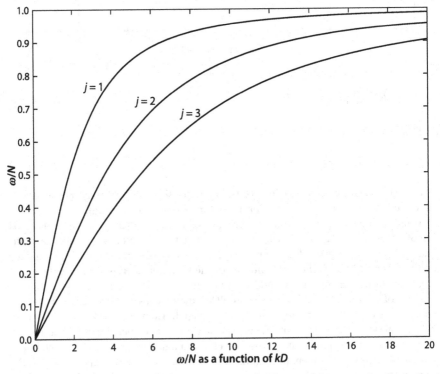

Fig. 9.6. The dispersion relation showing the frequency as a function of wave number for the first three internal gravity wave modes

The normal modes, of course, do not propagate energy vertically. Each mode in z can be decomposed into two plane waves using each with vertical wave number of opposite sign so that the eigenfunction can be thought of as the sum of an upward and downward propagating mode whose energy fluxes vertically cancel.

There is energy propagation in the horizontal direction, and for each vertical mode:

$$\frac{\partial \omega_j}{\partial k} = ND \frac{j\pi}{\left(j^2\pi^2 + k^2 D^2\right)^{3/2}} j = 1,2,3\ldots \tag{9.49}$$

Note that for very large kD, the group velocity in the horizontal direction goes to zero. The maximum group velocity for each mode is $c_{g\,max} = ND / j\pi$, which occurs for the long waves, i.e., as $kD \longrightarrow 0$.

Since N is a relatively strong function of z in the natural ocean, it is important to point out that *qualitatively* the eigenfunctions and eigenvalues of the more general problem are similar to the case of constant N. There are, however, a few important features of the solution to consider when N is variable.

Consider again the governing equation for the eigenfunction, $W(z)$:

$$\frac{d^2 W}{dz^2} + k^2 \left[\frac{N^2}{\omega^2} - 1\right] W = 0 \tag{9.50a}$$

$$W = 0, \quad z = -D, 0 \tag{9.50b}$$

It should be clear that the solutions of the above problem satisfying the homogeneous boundary conditions on W must have frequencies less than the maximum value of N in the interval $-D < z < 0$. This follows intuitively from the nature of the equation, for if the frequency is greater than N_{max}, the coefficient in front of the last term in the equation is always negative and the character of the solutions will be exponential rather than wavelike, and it will be impossible to satisfy the two homogeneous boundary conditions. This also follows from multiplication of the W equation by W, which with integration by parts and use of the boundary conditions yields

$$\int_{-D}^{0} \left[-\left(\frac{dW}{dz}\right)^2 + k^2\left(\frac{N^2}{\omega^2} - 1\right) W^2\right] dz = 0 \tag{9.51}$$

If $\omega > N$ everywhere in the depth interval, both terms in the integral would be negatively definite, and there would be no way to satisfy the condition that the integral vanish. Similarly, it follows from the theory of standard Sturm-Liouville problems that the eigenfunctions corresponding to different eigenvalues are orthogonal:

$$\int_{-D}^{0} W_j W_k N^2(z) dz = 0, \quad i \neq j \tag{9.52}$$

In addition, the eigenfunction corresponding to each higher eigenvalue has one more zero of the function $W(z)$ in the depth interval.

The character of the eigenfunctions are of interest. If

$$N_{min}^2 \leq \omega^2 \leq N_{max}^2 \tag{9.53}$$

the character of W will be oscillatory in the depth interval in which N is greater than ω and evanescent outside that interval. We can therefore expect some modes to have their energy trapped in the region where the stratification is greatest, and these will be the modes with the highest frequencies. There is a very good discussion of the general problem in Gill's book, and Fig. 9.7 and 9.8 are taken from that book.

Figure 9.7 shows a characteristic profile of N in the subtropical North Atlantic. Note the maximum of N in the region of the thermocline at about 750 meters. Figure 9.8 shows the first two eigenfunctions for that profile.

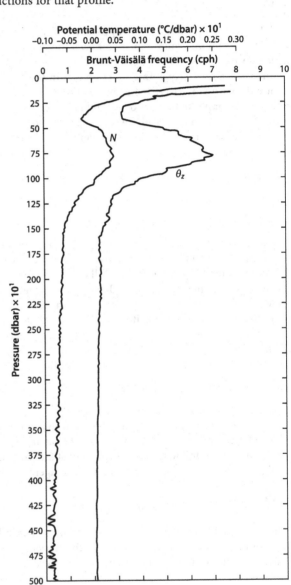

Fig. 9.7.
The distribution of $N(z)$ in the
North Atlantic (from Gill 1982)

The figures on the left of the figure are essentially W_j. The second figure is essentially the form of the solution in the long wave limit (more of this will be discussed later), and the last figure is the shape of the pressure or horizontal velocity in each mode, really the derivative of the function W. Note, as expected, the $n = 1$ mode has no zeros for W (just like $\sin \pi z / D$), while the second mode has a single zero (like $\sin 2\pi z / D$). However, the location of the zero and maxima differ from the constant N case. Note too that at great depth where N is small, the eigenfunctions are not oscillatory in agreement with out qualitative discussion of the governing equation.

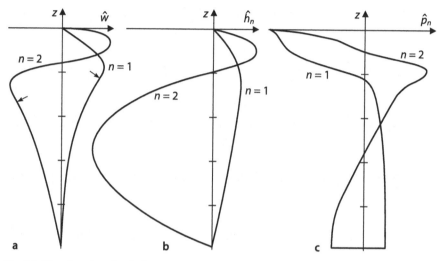

Fig. 9.8. Eigenfunctions for the buoyancy profile of Fig. 9.7 (from Gill 1982)

Vertical Propagation of Waves: Steady Flow and the Radiation Condition

There are numerous situations in which fluid flows over an obstacle, say a mountain in the atmosphere, a sea mount, or a ridge in the ocean, and we would imagine that internal gravity waves, if the fluid is stratified, would be generated. Such situations are of interest in their own right, but additionally they force us to carefully examine the radiative properties of the waves, which must be understood, sometimes, to actually solve the problem.

Consider the case of a stratified, incompressible fluid as studied in the preceding lectures, except now we will imagine that the background state includes a mean velocity in the x-direction, which is also a function of depth, z. If the dynamics are inviscid, such unidirectional flows are themselves exact solutions of the equations of motion. Indeed, consider the equation of motion in the zonal direction. In the absence of rotation and friction, and for motion in the x-z-plane,

$$\frac{\partial u}{\partial t} + u\frac{\partial u}{\partial x} + w\frac{\partial u}{\partial z} = -\frac{1}{\rho}\frac{\partial p}{\partial x} \tag{10.1}$$

Before we start to examine the wave problem, first note that if the flow is periodic in the x-direction, the x-*average* denoted by an overbar will yield

$$\frac{\partial \overline{u}}{\partial t} = -\frac{\partial \overline{uw}}{\partial z} \tag{10.2}$$

To obtain this equation, we have used the continuity equation and the assumption that the x-average of all terms of the form $\partial/\partial x$[finite] will be zero. The equation above describes how the mean current can change if there is a convergence of the flux of x-momentum carried by vertical motion $\overline{uw}$. Note that this term can be different from zero *even if u and w separately have zero x-average*. It is therefore possible for the waves that we will consider to alter the very flow that constitutes the character of the background in which the waves are embedded, but note that the change will be quadratic in the amplitude of the wave motions (if the flux $\overline{uw}$ is due to the waves). Thus, in linear theory for the waves, the background flow can be approximated by its initial value, but the above equation can be used to consistently calculate the small, second order changes in the mean due to the waves, and sometimes we are very interested in that change.

To linearize the equations of motion, we write all variables as a sum of the order one basic state plus a small perturbation. For our fields, that will be

$$u_{total} = U(z) + u(x,z,t) + \dots \tag{10.3a}$$

$$w_{total} = w(x,z,t) + \dots \tag{10.3b}$$

$$p_{total} = p_0(z) + p(x,z,t) + \dots \tag{10.3c}$$

$$\rho_{total} = \rho_0(z) + \rho(x,z,t) + \dots \tag{10.3d}$$

where the lower case variables are the wave fields, and each is the order of the wave amplitude and hence is small compared with the basic state. Again, the basic state density and pressure fields, denoted by a subscript 0 satisfy the hydrostatic relation.

The linearized equations of motion, when the above decomposition is inserted into the equations of motion and only terms that are linear in the wave amplitude are retained, are

$$\rho_0 \left[\frac{\partial}{\partial t} + U \frac{\partial}{\partial x} \right] u + \rho_0 w \frac{dU}{dz} = -\frac{\partial p}{\partial x} \tag{10.4a}$$

$$\rho_0 \left[\frac{\partial}{\partial t} + U \frac{\partial}{\partial x} \right] w + \rho g = -\frac{\partial p}{\partial z} \tag{10.4b}$$

$$\frac{\partial u}{\partial x} + \frac{\partial w}{\partial z} = 0 \tag{10.4c}$$

$$\left[\frac{\partial}{\partial t} + U \frac{\partial}{\partial x} \right] \rho + w \frac{\partial \rho_0}{\partial z} = 0 \tag{10.4d}$$

Once again, we have assumed that the motion is incompressible and adiabatic (Eq. 10.4d). In the continuity equation, we have assumed that the basic state density changes by a very small amount over the vertical scale of the motion. Except for the presence of terms proportional to $U(z)$, this set of equations is identical to the equations we used to study internal gravity waves. Note that the last equation could be rewritten in the form

$$\left[\frac{\partial}{\partial t} + U \frac{\partial}{\partial x} \right] b - w N^2 = 0 \tag{10.5a}$$

$$b \equiv g \rho / \rho_0 \tag{10.5b}$$

It is convenient to introduce a compact notation for the linearized advective operator, i.e.,

$$\tilde{D} \equiv \left[\frac{\partial}{\partial t} + U \frac{\partial}{\partial x} \right] \tag{10.6}$$

in terms of which our equations are

$$\rho_0 \tilde{D} u + \rho_0 w U_z = -p_x \tag{10.7a}$$

$$\rho_0 \tilde{D} w = -p_z - b\rho_0 \tag{10.7b}$$

$$u_x + w_z = 0 \tag{10.7c}$$

$$\tilde{D} b = w N^2 \tag{10.7d}$$

where, for ease, subscripts have been used to denote differentiation.

Note that using the linearized Lagrangian relation between vertical displacement ζ and the vertical velocity

$$\tilde{D}\zeta = w \tag{10.8}$$

the adiabatic equation yields

$$b = \zeta N^2 \tag{10.9}$$

Multiplying each momentum equation by its velocity component and the adiabatic equation by the buoyancy, b, with the aid of the continuity equation, we obtain the energy equation

$$\rho_0 \tilde{D} \left\{ \frac{u^2 + w^2}{2} + \frac{\zeta^2 N^2}{2} \right\} + \nabla \cdot p\vec{u} = -\rho_0 u w \frac{dU}{dz} \tag{10.10}$$

If the vertical shear of the basic state velocity U is zero, then we obtain the usual conservation statement for the sum of the kinetic and potential energies. However, when the shear is different from zero, there is a *source term for the perturbation wave energy*. This source is proportional to the shear and its product with the product $-\rho_0 u w$, *the Reynolds stress*. We are particularly interested in the sign of this term when averaged over a wave period or wavelength, $-\rho_0 \overline{uw}$. Consider the situation depicted in Fig. 10.1.

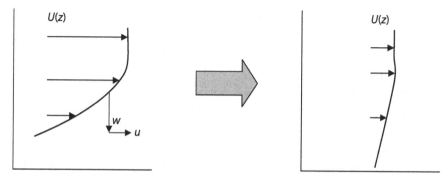

Fig. 10.1. Illustrating the flux of momentum in the Reynolds stress. The velocity profile is smoothed by the transfer of momentum by the waves

Think about a region in which $dU/dz > 0$. If, *on average*, the perturbation zonal velocity u is positive whenever w is negative, the source term will be positive and the energy of the wave field will tend to increase (of course the energy so produced could be locally fluxed away). Is that likely? If a fluid element has $w < 0$ with $dU/dz > 0$, then it is coming from a region where U is larger than where it arrives. If it retains to some degree its original x-momentum, it will show up at its new location with a perturbation u, which is positive. Now the "if" of the last sentence is a big one, since there is no guarantee that other factors, such as the perturbation pressure, might not intervene to alter that simple prediction. In fact, it often occurs that the Reynolds stress turns out to be zero even in the presence of shear and perturbations u and w. When that correlation $-\rho_0 \overline{uw}$ is different from zero, it provides a flux of u-momentum from one z-level to another. If that flux, as described above, is *downgradient*, i.e., from larger U to smaller U, the perturbation energy will increase. However, in fluxing mean momentum down the gradient it will tend to locally "flatten" the profile of mean velocity as shown in the figure. Now from elementary considerations, this internal mechanism cannot alter the overall mean momentum of the flow; that is,

$$\int_{\text{all } z} U dz = \text{constant} \tag{10.11}$$

However, the mean kinetic energy is

$$\int_{\text{all } z} U^2 \rho_0 dz / 2$$

It is easy to see that if the integral of U is fixed and the profile of U is made more flat, the variance of U, i.e., the kinetic energy of the mean flow will decrease.

This can also be seen directly by considering the momentum equation for the mean flow in x as written above:

$$\frac{\partial \overline{u}}{\partial t} = -\frac{\partial \overline{uw}}{\partial z} \tag{10.12}$$

Multiplication by $\overline{u}$ yields

$$\frac{1}{2} \frac{\partial \overline{u}^2}{\partial t} = -\overline{u} \frac{\partial \overline{uw}}{\partial z} = -\frac{\partial \overline{u}\,\overline{uw}}{\partial z} + \overline{uw} \frac{\partial \overline{u}}{\partial z} \tag{10.13a}$$

or

$$\frac{1}{2} \frac{\partial \overline{u}^2}{\partial t} + \frac{\partial \overline{u}\,\overline{uw}}{\partial z} = \overline{uw} \frac{\partial \overline{u}}{\partial z} \tag{10.13b}$$

The first term in the above equation is the rate of change of the kinetic energy associated with the mean flow (per unit mass). The second term on the left is a flux term, which will integrate to zero if the flow is contained between horizontal plates where w vanishes. The term on the right-hand side is a source or *sink* of kinetic energy of the mean. Comparing it to the equation for the perturbation energy we see that if the Reynolds stresses increase the energy of the perturbation wave field, they must at the same time be decreasing the energy of the mean flow. (Note in our identification we are using the fact that to order amplitude squared, $\bar{u} = U$.) Thus, this term is an *energy transformation term* representing an energy transfer between the mean (over a period or wavelength) and the perturbations.

The vertical flux of wave energy is again $\overline{pw}$. What can we say about it and its relation to the Reynolds stress? The earliest treatment of the problem can be found in Eliassen and Palm (1960) and the student is referred to it for an fuller understanding of the historical context.

We are going to be especially interested in steady flows over bumps and the resulting steady wave field generated by the interaction of the flow and the topography.

In the steady state in which there is no secular increase in the wave energy, the energy equation reduces, when averaged in x, to

$$\frac{\partial}{\partial z}\{\overline{pw} + \rho_0 \overline{uw}U\} = \rho_0 U \frac{\partial \overline{uw}}{\partial z} \tag{10.14}$$

after a simple integration by parts of the source term.

On the other hand, the steady momentum equation in the x-direction is

$$\frac{\partial}{\partial x}[\rho_0 Uu + p] + \rho_0 w \frac{dU}{dz} = 0 \tag{10.15}$$

Since the motion is two-dimensional and nondivergent, we can introduce a stream function ψ such that

$$u = -\psi_z \tag{10.16a}$$

$$w = \psi_x \tag{10.16b}$$

which allows the momentum equation to be rewritten as

$$\frac{\partial[\rho_0 Uu + p + \rho_0 \psi U_z]}{\partial x} = 0 \tag{10.17}$$

If the motion is periodic in x or if it vanishes as $x \longrightarrow$ infinity, then the quantity inside the square bracket must itself vanish, so that

$$[\rho_0 Uu + p + \rho_0 \psi U_z] = 0 \tag{10.18}$$

If this equation is multiplied by w,

$$\rho_0 uwU + pw = -\frac{1}{2}\frac{\partial \psi^2}{\partial x}\rho_0 \frac{dU}{dz} \tag{10.19}$$

where the relation between ψ and w has been used. An average over a wavelength in x then yields the important result:

$$\overline{pw} + \rho_0 U \overline{uw} = 0 \tag{10.20}$$

From this it follows from the energy equation that

$$\frac{\partial \overline{uw}}{\partial z} = 0 \tag{10.21}$$

This implies that even if $\overline{uw} \neq 0$, its derivative with z *must be zero if the wave field is steady and there is no dissipation*. Returning to the first equation of this lecture for the mean flow, it also means that the wave field will not, under these circumstances, alter the mean. This will occur only for that part of the wave field that has its quadratic properties varying with time, for example at the front of an otherwise steady wave field.

The relationship of the vertical energy flux and the Reynolds stress allows a simple interpretive tool to characterize the sign of the energy flux. From $\overline{pw} + \rho_0 \overline{uw} = 0$, it follows that

$$\overline{pw} = -\rho_0 U \overline{uw} = \rho_0 U \overline{\psi_x \psi_z}$$

$$= \rho_0 U \overline{\frac{\psi_x}{\psi_z} \psi_z^2} = -\rho_0 U \overline{\left(\frac{\partial z}{\partial x}\right)_\psi u^2} \tag{10.22}$$

If the vertical energy flux is positive, this implies that the slope in the x-z-plane of the lines of constant ψ (these are the phase lines of the wave) must be negative if U is positive (Fig. 10.2).

Now let's derive the governing equation for *steady* perturbations. By taking the z-derivative of the x-momentum equation and subtracting from that the x-derivative of the vertical momentum equation, we obtain

$$\rho_0 \tilde{D}[w_x - u_z] - \rho_0 U_z u_x - \rho_0 w_z U_z - \rho_0 w U_{zz} = -\rho_0 b_x \tag{10.23}$$

which is the equation for the y-component of vorticity. The term on the right-hand side represents the baroclinic production of relative vorticity by horizontal density gradients in the wave (this is the linearized part of $\nabla p \times \nabla \rho$). With the aid of the continuity equation, this becomes

$$\tilde{D}\nabla^2 \psi - \psi_x U_{zz} = -b_x \tag{10.24}$$

while the adiabatic equation is

$$\tilde{D}b = N^2 \psi_x \tag{10.25}$$

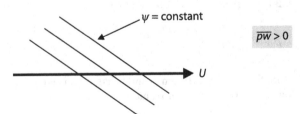

$\psi = $ constant

$\overline{pw} > 0$

U

Fig. 10.2.
The orientation of wave crests
to yield upward wave radiation

Eliminating b between the two equations yields the final equation for the stream function:

$$\tilde{D}^2 \nabla^2 \psi + N^2 \psi_{xx} = \tilde{D}\psi_x U_{zz} \tag{10.26}$$

We note that if $U = 0$, the equation reduces to the equation for internal gravity waves we obtained in the last lecture, since now $\tilde{D} = \partial/\partial t$ if U is zero.

However, we are interested in the case where U is not zero but where the flow and the wave field are steady. In the steady state, $\tilde{D} = U\partial/\partial x$, and so the governing equation for ψ can be written in the compact form

$$U^2 \frac{\partial^2}{\partial x^2}\left[\nabla^2 \psi + \left(\frac{N^2}{U^2} - \frac{U_{zz}}{U}\right)\psi\right] = 0 \tag{10.27}$$

For motion that is periodic in x (or which vanishes at infinity) we can integrate the above equation twice to obtain

$$\nabla^2 \psi + \left(\frac{N^2}{U^2} - \frac{U_{zz}}{U}\right)\psi = 0 \tag{10.28}$$

For solutions that are periodic in x with wave number k, we can look for solutions of the form

$$\psi = \phi(z)e^{ikx} \tag{10.29}$$

where it is understood that we take the real part of the solution.

Thus, ϕ satisfies

$$\frac{d^2\phi}{dz^2} + (\ell^2(z) - k^2)\phi = 0 \tag{10.30a}$$

$$\ell^2 \equiv \left\{\frac{N^2}{U^2} - \frac{U_{zz}}{U}\right\} \tag{10.30b}$$

A particularly illuminating example occurs when we consider the flow over a bumpy lower boundary, whose elevation is given by the periodic form

$$h = h_0 \cos kx$$

(which is why we chose k for the wave number of our solution).

The lower boundary condition is

$$w = \frac{dh}{dt} = U\frac{\partial h}{\partial x} \tag{10.31}$$

where the last equality depends on the solution being steady and linear.

We will imagine that the upper boundary is very far away and idealize that by considering that z runs between 0 at the lower boundary and ∞ for large positive z.

Since $w = \partial \psi / \partial x$, an integration of the lower boundary condition in x yields

$$\psi = Uh , \quad z = 0 \tag{10.32a}$$

or

$$\psi = Uh_0 \operatorname{Re} e^{ikx} \;\Rightarrow\; \phi = Uh_0 , \quad z = 0 \tag{10.32b}$$

A glance at the differential equation for ϕ shows that the character of the solution depends on whether

$$\ell^2 > k^2$$

or

$$\ell^2 < k^2$$

In the former case, the solutions will be oscillatory in z, while in the latter case they will be exponential in z. For the simple case of constant U, this change in character has a simple physical interpretation. In this case, $\ell^2 = N^2 / U^2$. In the frame in which the lower boundary is fixed and the flow is moving to the right with speed U, the motion is steady. Let's put ourselves in a frame moving to the right with the basic flow. Then the unperturbed fluid appears to be at rest, but it is being disturbed by a lower boundary with a ripple of wave number k moving to the right at speed U. This will force a response with the forcing frequency of the boundary disturbance, which is Uk. If that frequency is greater than the maximum internal gravity wave frequency, N, we clearly can't have a wavelike response in z. The condition that $Uk < N$ is simply $\ell^2 = N^2 / U^2 > k^2$.

Let us first examine the other case where $\ell^2 < k^2$.

In this case, the solution for ϕ is

$$\phi = Ae^{-mz} + Be^{mz}$$
$$m = \left[k^2 - \ell^2\right]^{1/2} > 0 \tag{10.33}$$

Now, for large positive z, we would like the solution to remain bounded, i.e., as $z \longrightarrow \infty$, we want ϕ to remain finite. This clearly requires that we choose $B = 0$. The remaining constant is determined by the condition at $z = 0$, namely,

$$A = Uh_0 \tag{10.34}$$

Thus,

$$\psi = \operatorname{Re} Uh_0 e^{-mz} e^{ikx} = Uh_0 e^{-mz} \cos kx = Ue^{-mz} h(x) \tag{10.35}$$

from which we obtain

$$w = \psi_x = -Uh_0 k e^{-mz} \sin kx \tag{10.36a}$$

$$u = -\psi_z = -Uh_0 m e^{-mz} \cos kx \tag{10.36b}$$

$$\overline{\psi_x \psi_z} = U^2 h_0^2 mk e^{-2mz} \overline{\sin kx \cos kx} = 0 \tag{10.36c}$$

Note that the streamlines are in phase with the topography and simply diminish exponentially with height above the bottom. Furthermore, the last equation tells us that the Reynolds stress and hence the vertical energy flux are *identically zero when averaged over a wave period*. That seems reasonable. In this parameter regime, no internal gravity wave can be excited (the frequency is too large), and without a wave response there is no upward radiation of energy. Note also that the vertical displacement

$$\zeta = \frac{\psi}{U} = h_0 e^{-mz} \cos kx = h(x) e^{-mz} \tag{10.37}$$

is exactly in phase with the topography. Where the bottom goes up, the streamline follows it.

Let's calculate the drag on the mountain by the pressure in the wave field. For a bottom with a relief $h = h(x)$,

$$\text{Drag} = \int_0^{2\pi/k} p \frac{\partial h}{\partial x} dx \tag{10.38}$$

That is, the drag is the pressure times the projection of the topography that presents a face perpendicular to the x-axis. There will be a drag if there is higher pressure on the face of the slope upstream compared to the pressure on the face downstream. We can easily calculate the pressure from the steady zonal momentum equation when $dU/dz = 0$. In that case,

$$p_x = -\rho_0 U u_x \quad \Rightarrow \quad p = -\rho_0 U u = \rho_0 U \psi_z \tag{10.39}$$

from which it follows in the present case,

$$p = -\rho_0 U^2 m h(x) e^{-mz} \tag{10.40}$$

Note that for this case, the pressure is in phase with the topography. The lowest pressure (the largest negative anomaly) occurs over the ridge crest, and the pressure is symmetric about the crest; indeed,

$$\text{Drag} = \int_0^{2\pi/k} p(x,0) h_x dx = -\rho_0 U^2 m \int_0^{2\pi/k} h h_x dx = 0 \tag{10.41}$$

Thus, in this parameter regime there is no wave radiation and no drag on the topography. The absence of drag is not too surprising. We know that for a homogeneous, incompressible, irrotational flow in the absence of friction there is no drag. Clearly, if Uk is much greater than N, the fluid will respond as if the stratification were zero and the zero drag result of potential flow is anticipated. It is perhaps a little strange that this result holds up to the equality (at least) where $Uk = N$, but as we shall see, the drag is directly related to the ability of the flow to support a wave, and the threshold $Uk = N$ is precisely that boundary between wave and no wave.

Now let's consider the more interesting case when $Uk < N$. This puts the fluid in the parameter regime in which internal gravity waves can be generated. In this case, we define

$$m^2 = \ell^2 - k^2 = \frac{N^2}{U^2} - k^2 > 0 \tag{10.42}$$

so that the solution for the wave is

$$\phi(z) = Ae^{imz} + Be^{-imz} \tag{10.43}$$

We still have the boundary condition at $z = 0$ that

$$\phi = Uh_0 , \quad z = 0 \tag{10.44}$$

However, the condition that the solution be finite at infinity is no help at all in rejecting either the A or B solution for ϕ, so that with the boundary condition at $z = 0$, we will have one condition (equation) for the two unknowns, A and B. How did we get into this pickle?

The essence of the difficulty is related to the two infinities we have introduced into our problem by our simplifications. First, we have assumed that since the upper boundary is so far away, we may idealize the region as infinite in z. Of course, for waves radiating upward, that will hold only for a finite time. Second, we have decided to examine the steady problem after all transients have radiated away and that requires that in principle, an infinite amount of time has passed so there is clearly a conflict between the two assumptions. Had we solved the initial value problem, i.e., if we had considered the problem for finite t while the z-domain was infinite, it would be clear that since the group velocity is finite, for *all finite t the disturbance should go to zero as z goes to infinity*. However, in the problem what we have done is to invert this limit. By examining the steady problem, we let time go to infinity first and then we must ask how the solution looks for large z. Stated this way, it is clear that we are asking for the partial solution that is valid within the advancing front that was set up long ago by the initial disturbance. Inside that front the solution may be steady, but without considering the initial value problem, there seems no way, with the boundary conditions so far applied, to specify the steady solution (find A and B). Because we have chosen to solve a simple, *physically incomplete* problem, we must add some physics to take the place of the initial value problem we have chosen not to solve (because it would be so complicated). Indeed, it would be a pity to have to go through the whole initial value problem just to determine which combination of A and B is correct in the steady state.

The physics that we must add is called *the radiation condition*. Simply put, it states that we must decide on the direction of the wave radiation in the steady state that is physically pertinent for our problem. If we consider the problem as one in which a disturbance is formed at the topography by its interaction with the current and then radiates upward, the condition should be that the radiation be *outgoing radiation*; that is, the energy flux in the z-direction should be positive.

We have seen before that the vertical energy flux is

$$\overline{pw} = \rho_0 U \overline{\psi_x \psi_z} \tag{10.45}$$

where again an overbar denotes an average in x over a wavelength.

With the above solution for ϕ, we have

$$\psi = A e^{i(kx+mz)} + B e^{i(kx-mz)} \tag{10.46}$$

The part of the solution with coefficient A has its phase lines tilting upstream (for $U > 0$), while the B solution has its phase lines tilting downstream (Fig. 10.3).

We saw earlier that the energy flux would be upward for phase lines tilting upstream so we would expect the A solution to be the one that gives us the physically acceptable solution of outgoing radiation. Let's calculate the energy flux explicitly.

With $\overline{pw} = \rho_0 U \overline{\psi_x \psi_z}$, we first calculate the energy flux in the A solution. Recall that we must use the *real* part of the solution. Thus, using * to denote complex conjugation,

$$\psi = \frac{A}{2} e^{i(kx+mz)} + \frac{A^*}{2} e^{-i(kx+mz)} \tag{10.47}$$

so that with $\theta \equiv kx + mz$

$$\overline{pw} = \rho_0 U \overline{\psi_x \psi_z} = \frac{i k \rho_0 U}{2} \overline{\left[A^{i\theta} + A^* e^{-i\theta} \right] \frac{im}{2} \left[A e^{i\theta} - A^* e^{-i\theta} \right]}$$

$$= \frac{-m k \rho_0 U}{4} \overline{\left[-2|A|^2 + A^2 e^{2i\theta} + A^{*2} e^{-2i\theta} \right]} \tag{10.48}$$

$$= \frac{mk}{2} \rho_0 U |A|^2 > 0$$

It is important to note that in calculating the quadratic product of pw, we had to use the real part of ψ, which involves both the linear solution and its complex conjugate. Had we erroneously used only the term proportional to $e^{i\theta}$, we would have obtained only terms like $e^{2i\theta}$ from the product, and these have *zero average* over a wavelength. The correct answer given above, whose average is different from zero, depends on using the full real parts to calculate the flux terms.

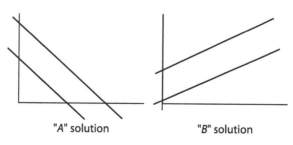

Fig. 10.3.
The tilt of the wave crests in the two solutions. The A solution has energy propagating upward; the B solution has energy propagating downward

"A" solution   "B" solution

If we were to make the same calculation with the B solution, the result would be

$$\overline{pw} = -\frac{mk}{2}\rho_0 U |B|^2 < 0 \tag{10.49}$$

Since by convention both m and k are positive, it is the A solution that represents outgoing radiation, and the B solution represents incoming radiation from infinity. Consistent with our physical description, it is the A solution which is appropriate to our problem, since that is the solution that satisfies the *radiation condition of outgoing energy flux*. It is important to realize that the B solution is a perfectly fine solution physically. It represents, in the context of the steady problem, wave energy moving downward from some source far away and above the topography. There may indeed be problems in which that radiation would be appropriate, but our specification of outward radiation is the additional condition that we must add to render our solution both unique and appropriate to the physical problem we have in mind.

Using the condition that $z = 0$ yields $A = Uh_0$ so that

$$\psi = \mathrm{Re}\, A e^{i(kx+mz)} = Uh_0 \cos(kx+mz) \tag{10.50}$$

this yields, as expected, phase lines tilting upstream (Fig. 10.4).

From the earlier result (Eq. 10.39),

$$p = -\rho_0 U u = \rho_0 U \psi_z = -\rho_0 U h_0 m \sin(kx+mz) \tag{10.51}$$

Thus on $z = 0$, $p = -\rho_0 U h_0 m \sin kx$, which is now 90° out of phase with the topography. High pressure now occurs on the upstream face of each ridge (where x is $-\pi/2$) while the downstream face has low pressure. This leads to a net force on the topography so that (Fig. 10.5)

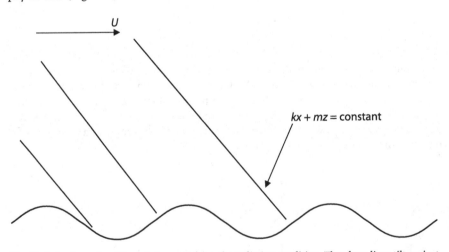

Fig. 10.4. A schematic of the solution satisfying the radiation condition. The phase lines tilt against the current

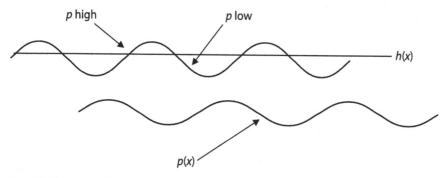

Fig. 10.5. The pressure distribution with respect to the topography for the case where waves are radiated

$$\text{Drag} = \int_{0}^{2\pi/k} p\,\frac{\partial h}{\partial x}\,dx = \rho_0 U^2 h_0^2\,\frac{km}{2} \tag{10.52}$$

Note that for $km > 0$, the correct choice of the phase orientation, the drag is positive. This drag has nothing to do with friction; it is simply the response to the wave energy that is radiated away to infinity by the topography. Indeed, in the case where there was no wave radiation, there was no drag. Note the relation between the wave drag on the topography and the flux of energy; using our result for the amplitude A,

$$\overline{pw} = \rho_0 U^3 h_0^2\,\frac{mk}{2} = \text{Drag}\times U \tag{10.53}$$

so that the rate at which energy is radiated away from the topography is precisely equal to the rate at which the drag is doing work on the topography and thus is equal to the rate at which the topography is doing work on the fluid.

There have been other techniques introduced to deal with the apparent indeterminacy of the solution for which we have used the radiation condition. An alternative is to introduce a small amount of friction, for ease, proportional to the velocity, and recalculate the constant m. With the presence of friction, m will be complex and one solution will exponentially increase with z, while the other decreases. Choosing the solution with the exponential decrease and then letting the size of the friction go to zero reproduces the solution found here by the radiation condition. The student should think through the physical reason for why this is true, and those with a background that includes the Laplace transform should also see why it is equivalent to the steady state chosen by the initial value problem.

We noted above that the choice of solution by the radiation condition is equivalent to choosing the solution that has the energy flux upwards. Let's spend a moment reviewing how that would enter explicitly in the steady problem.

Let prime variables denote velocities and positions seen by an observer moving to the right at the speed U. For such an observer, there will be no mean flow and as already mentioned, that observer will see a rippled lower boundary moving to the left with speed $-U$ forcing the fluid with a frequency Uk.

To move into the primed frame, we introduce the relations

$$x = x' + Ut'$$
$$t = t'$$

(10.54)

so that the phase in the resting frame $e^{i(kx-\omega t)}$ becomes $e^{i(kx'+[kU-\omega]t')}$ in the moving frame. We define the *intrinsic frequency* as the frequency seen in the frame moving with the mean flow so that it is the frequency with respect to a locally stationary fluid:

$$\omega_0 = \omega - kU \quad \text{or}$$

(10.55a)

$$\omega = Uk + \omega_0$$

(10.55b)

Thus, in the original frame in which the lower boundary is at rest, the frequency is the sum of the intrinsic frequency plus the *Doppler shift Uk*. We know that the intrinsic frequency for internal gravity waves, i.e., in the absence of a mean flow, is

$$\omega_0 = \pm \frac{Nk}{\sqrt{k^2 + m^2}}$$

(10.56)

In the case under consideration, we are taking into account a disturbance which is *steady* in the rest frame so that the frequency ω must be zero. This means we must choose the negative sign in the above equation for the intrinsic frequency so that the Doppler shift downstream cancels the phase propagation upstream.

The condition of zero frequency chooses the wave number component m such that

$$m^2 = \frac{N^2}{U^2} - k^2$$

a result already obtained from our differential equation for ϕ. Although the frequency is zero, its derivative with respect to wave number is different from zero. Therefore the group velocity in the resting frame is

$$c_{gx} = \frac{\partial \omega}{\partial k} = U - \frac{Nm^2}{\left(k^2 + m^2\right)^{3/2}}$$

(10.57a)

$$c_{gz} = \frac{Nkm}{\left(k^2 + m^2\right)^{3/2}}$$

(10.57b)

Using the above results for m yields

$$c_{gx} = \frac{Uk^2}{K^2}$$

(10.58a)

$$c_{gz} = \frac{Ukm}{K^2}$$

(10.58b)

so that the direction of the group velocity is *downstream* and upwards even though the phase lines are tilting upstream.

Thus,

$$\frac{c_{gz}}{c_{gx}} = \frac{m}{k} \tag{10.59}$$

so that $\vec{c}_g$ is *parallel* to the wave vector $\vec{K}$. In the frame in which there is no mean flow, our earlier results (or a recalculation using the intrinsic frequency) yield

$$c_{gx_0} = -\frac{Nm^2}{K^{3/2}} \tag{10.60a}$$

$$c_{gz_0} = \frac{Nkm}{K^{3/2}} \tag{10.60b}$$

so that in the frame of no mean flow, as previously noted, the group velocity is at right angles to the wave vector

$$\frac{c_{gz_0}}{c_{gx_0}} = -\frac{k}{m} \tag{10.61}$$

The relation between the two is illustrated in Fig. 10.6. It is a simple matter to show that the two group velocities are orthogonal to one another; that is, $\vec{c}_g \cdot \vec{c}_{g0} = 0$.

The difference between the two group velocities is precisely equal to the mean flow in the x-direction U, which carries energy of the wave field downstream. The two wave vectors and their relation to the mean flow are shown in Fig. 10.7. A very clear discussion of the relationship is to be found in Lighthill (1978).

A final remark is in order about the relation between the drag and the effect on the mean flow. Since the drag represents a force on the fluid by the topography, one ex-

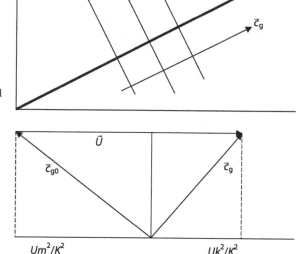

Fig. 10.6.
The wave vector and the direction of the group velocity for the steady, radiated internal gravity wave

Fig. 10.7.
The relationship between the group velocity vector for the steady wave and the wave in a resting medium

pects that the mean flow will decelerate as a consequence. However, we earlier noted that for steady waves in the absence of dissipation,

$$\frac{\partial \overline{uw}}{\partial z} = 0 \quad \text{and since}$$

$$\frac{\partial \overline{u}}{\partial t} = -\frac{\partial \overline{uw}}{\partial z}$$

where the right-hand side is calculated from the wave field, there is apparently no deceleration of the mean flow. How is this apparent paradox resolved? (Note: usually a paradox is a sign of incomplete thought, not an out and out error).

As mentioned above in our discussion of the radiation condition, there will be an outgoing front, ahead of which there is no wave signal, behind which we will have established the steady wave field of our calculation. At a time t, the front will have moved to a distance $c_{gz}t$ above the topography into the fluid. At the position of the front, the above momentum balance holds. So ahead of the front $\overline{uw} = 0$ while behind the front (Fig.10.8)

$$\overline{uw} = U^2 h_0^2 \frac{mk}{2}$$

Thus,

$$-\frac{\partial \overline{uw}}{\partial z} < 0$$

and is large at the front. It is at the front that the deceleration occurs, and only there. Integrating across the front yields a local change in the mean

$$\Delta \overline{u} = -\frac{U^2 h_0^2 mk}{2c_{gz}} = -Uh_0^2 K^2 / 2$$

Note that the change is of the order of the small parameter $(h_0 K)^2$, which is the square of the steepness of the topography.

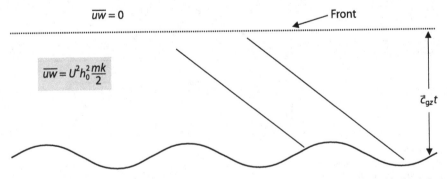

Fig. 10.8. The waves radiated from the topography and the Reynolds stress developed behind the advancing front

Rotation and Potential Vorticity

For motions whose time scales are of the order of a day or greater, or more precisely when the frequency of the wave motion is of the order of the Coriolis parameter or less, the effects of the Earth's rotation can no longer be ignored. Such waves are evident in both oceanic and atmospheric observational spectra. Figure 11.1 taken from the article of Garrett and Munk (1979) shows a power spectrum of vertical displacement of an isotherm. We see a great deal of variance at frequencies less than N (as we might expect) with a peak near the Coriolis frequency $f = 2\Omega \sin \theta$.

Some waves, such as gravity waves, are *affected* by rotation while others are primarily *due* to rotation, and of these there are different types with different characteristic time scales.

Consider first an unbounded fluid. To simplify the analysis, we will start by assuming that it is incompressible, inviscid and uniformly rotating. We will also assume that the vertical scale of the motion is much smaller than the scale over which the density

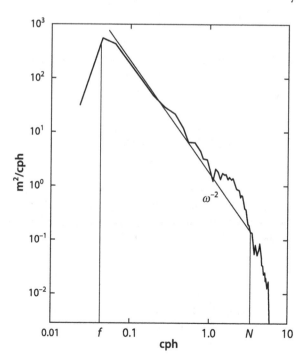

Fig. 11.1.
A power spectrum of internal wave energy (after Cairns and Williams 1976)

would change by an $O(1)$ amount, i.e., we assume that if m is the vertical wave number of the motion,

$$\frac{1}{\rho_0}\frac{\partial\rho_0}{\partial z} \ll m \tag{11.1}$$

For small perturbations we write, as before,

$$\rho_{total} = \rho_0 + \rho, \quad \rho \ll \rho_0 \tag{11.2a}$$

$$p_{total} = p_0 + p, \quad p \ll p_0 \tag{11.2b}$$

$$\frac{\partial p_0}{\partial z} = -\rho_0 g \tag{11.2c}$$

Then the linearized equations of motion for a fluid whose rotation axis is anti-parallel to the direction of gravity are

$$\rho_0\left[\frac{\partial u}{\partial t} - fv\right] = -\frac{\partial p}{\partial x} \tag{11.3a}$$

$$\rho_0\left[\frac{\partial v}{\partial t} + fu\right] = -\frac{\partial p}{\partial y} \tag{11.3b}$$

$$\rho_0\frac{\partial w}{\partial t} = -\frac{\partial p}{\partial z} - \rho g \tag{11.3c}$$

$$\frac{\partial u}{\partial x} + \frac{\partial v}{\partial y} + \frac{\partial w}{\partial z} = 0 \tag{11.3d}$$

$$\frac{\partial b}{\partial t} - wN^2 = 0 \tag{11.3e}$$

$$b = g\rho/\rho_0, \quad N^2 = -\frac{g}{\rho_0}\frac{\partial\rho_0}{\partial z} \tag{11.3f}$$

Our goal is to derive a single equation for either the pressure or vertical velocity to serve as our wave equation. To start, we take the x-derivative of the y-momentum equation and subtract from that the y-derivative of the x-momentum equation; that is, we are taking the curl of the horizontal momentum equations to obtain

$$\frac{\partial\zeta}{\partial t} = -f(u_x + v_y) = f\frac{\partial w}{\partial z} \tag{11.4}$$

Note that the Coriolis parameter in this study is assumed independent of position, an assumption that will be relaxed later in the course. The component of vorticity parallel to the z-axis is

$$\zeta = v_x - u_y \tag{11.5}$$

It is important to note that in the presence of the background rotation, z-dependent vertical motion will give rise to relative vorticity by stretching the vortex lines associated with the overall rotation of the fluid. If instead of taking the curl of the horizontal momentum equations we take the divergence, i.e., the x-derivative of the x-momentum equation and the y-derivative of the y-momentum equation, we obtain

$$\frac{\partial}{\partial t}\left(u_x + v_y\right) - f\zeta = -\frac{\nabla_h^2 p}{\rho_0} \tag{11.6a}$$

or

$$\frac{\partial^2 w}{\partial t \partial z} + f\zeta = \frac{\nabla_h^2 p}{\rho_0} \tag{11.6b}$$

If we eliminate w between the vorticity and the divergence equation,

$$\frac{\partial^2 \zeta}{\partial t^2} + f^2\zeta = f\frac{\nabla_h^2 p}{\rho_0} \tag{11.7}$$

while eliminating the vorticity between the same two equations yields

$$\frac{\partial^2}{\partial t^2}\frac{\partial w}{\partial z} + f^2\frac{\partial w}{\partial z} = f\frac{\nabla_h^2 p}{\rho_0} \tag{11.8}$$

If the perturbation is independent of horizontal position, the right-hand side of the above equation would be identically zero, and that would give rise to a harmonic oscillation at the Coriolis frequency f. This is analogous to the situation we saw for a nonrotating fluid in which disturbances independent of z gave rise to oscillations with frequency N. In the latter case, we discovered N as a limiting frequency of oscillation. We shall discover the same thing in the presence of rotation for oscillations with frequency f.

If we take the time derivative of the vertical momentum equation and use the adiabatic equation to eliminate the buoyancy b,

$$\frac{\partial^2 w}{\partial t^2} + N^2 w = -\frac{1}{\rho_0}\frac{\partial^2 p}{\partial t \partial z} \tag{11.9}$$

so that again, disturbances independent of z will oscillate with frequency N. We now have two limiting cases to keep an eye on. If we eliminate the pressure between Eq. 11.1 and Eq. 11.2, we obtain a single equation for w (take the horizontal Laplacian of Eq. 11.2 and the vertical derivative of Eq. 11.1):

$$\frac{\partial^2}{\partial t^2}\left[\nabla^2 w\right] + f^2\frac{\partial^2 w}{\partial z^2} + N^2\nabla_h^2 w = 0 \tag{11.10}$$

When f is zero, this recovers the wave equation for internal gravity waves studied earlier. To derive this equation, we have used repeatedly the approximation that the vertical derivative of the background density is small with respect to the vertical derivative of w, i.e.,

$$\frac{1}{\rho_0}\frac{\partial\rho_0}{\partial z} \ll \frac{1}{w}\frac{\partial w}{\partial z} \tag{11.12}$$

Before proceeding to the solution of the wave equation for w, it is illuminating to examine the equations of motion a bit more carefully.

First of all, note that if we knew the pressure, we could easily find the horizontal velocity $\vec{u}_h$ from the easily derived equation

$$\frac{\partial^2 \vec{u}_h}{\partial t^2} + f^2\vec{u}_h = -\frac{1}{\rho_0}\frac{\partial}{\partial t}\nabla_h p + \frac{f}{\rho_0}\hat{z}\times\nabla p \tag{11.13}$$

where $\hat{z}$ is the unit vector in the z-direction parallel to the rotation (and anti-parallel to gravity).

Second, we can find an equation for the pressure by eliminating w between Eq. 11.8 and Eq. 11.9 instead of the other way around. We obtain

$$\frac{\partial}{\partial t}\left[\frac{\partial^2}{\partial t^2}\nabla^2 p + f^2\frac{\partial^2 p}{\partial z^2} + N^2\nabla_h^2 p\right] = 0 \tag{11.14}$$

This is *almost* the same equation we obtained for w. There is an extra time derivative operating on the whole equation. For motions of nonzero frequency that would make no difference, but we must be careful. A first integral of the equation yields

$$\frac{\partial^2}{\partial t^2}\nabla^2 p + f^2\frac{\partial^2 p}{\partial z^2} + N^2\nabla_h^2 p = \Omega(x,y,z) \tag{11.15}$$

The question is what is Ω? Its existence is connected with the *conservation equation*, whose first integral is the above equation. It is also clear that it ought to be determined by initial data since it is independent of time. Clearly, then, it should be related to some quantity that is conserved during the motion.

If we return to the vorticity equation

$$\frac{\partial\zeta}{\partial t} = f\frac{\partial w}{\partial z} \tag{11.16}$$

and use the adiabatic equation to eliminate w, we obtain

$$\frac{\partial}{\partial t}\left(\zeta - f\frac{\partial(b/N^2)}{\partial z}\right) = 0 \tag{11.17}$$

This is, for the simple model we are considering, the form of the conservation of *potential vorticity q*, where

$$q = \zeta - f \frac{\partial b / N^2}{\partial z} \tag{11.18}$$

this form can be easily interpreted. Let's call (to avoid confusion with the symbol for vertical relative vorticity) the vertical *displacement* of a fluid parcel, Z. From the adiabatic equation, $Z = b / N^2$ so that

$$q = \zeta - f \frac{\partial Z}{\partial z} \tag{11.19}$$

If the potential vorticity (pv) is conserved, the spreading apart of density surface, i.e., $\partial Z / \partial z > 0$ in the presence of background planetary vorticity, f, will give rise to a corresponding increase in relative vorticity to keep q constant. This should be familiar from simple layer model treatments of pv.

It seems likely that our conserved quantity Ω is related to q. Can we show what the relation is? To simplify the derivation, we will take N constant.

From

$$f\zeta = \frac{\nabla_h^2 p}{\rho_0} - w_{zt} \tag{11.20a}$$

$$w_{zt} = b_{zzt} / N^2 \tag{11.20b}$$

$$\Rightarrow f\zeta = \frac{\nabla_h^2 p}{\rho_0} - \frac{1}{N^2} \frac{\partial}{\partial z} \frac{\partial^2}{\partial t^2} \left(-\frac{p_z}{\rho_0} - w_t \right) \tag{11.20c}$$

$$= \frac{\nabla_h^2 p}{\rho_0} + \frac{1}{N^2} \frac{p_{zztt}}{\rho_0} + \frac{1}{N^2} w_{zttt} \tag{11.20d}$$

and

$$-f^2 b_z / N^2 = \frac{-f^2}{N^2} (-p_{zz} - w_{zt}) \tag{11.21}$$

it follows that

$$f\left\{ \zeta - f \frac{b_z}{N^2} \right\} = \nabla_h^2 p / \rho_0 + \frac{1}{N^2 \rho_0} p_{zztt} + \frac{1}{N^2} w_{zttt} + \frac{f^2}{N^2} p_{zz} / \rho_0 + \frac{f^2}{N^2} w_{zt} \quad \text{or} \tag{11.22}$$

$$fq = \nabla_h^2 p / \rho_0 + \frac{f^2}{N^2} p_{zz} / \rho_0 + \frac{1}{N^2} \frac{\partial}{\partial t} \left(w_{ttz} + f^2 w_z \right)$$

$$= \nabla_h^2 p / \rho_0 + \frac{f^2}{N^2} p_{zz} / \rho_0 + \frac{1}{N^2} \frac{\partial^2}{\partial t^2} \left(\nabla_h^2 p / \rho_0 \right) \tag{11.23}$$

$$= \frac{1}{\rho_0 N^2} \left[\frac{\partial^2 \nabla^2 p}{\partial t^2} + f^2 \frac{\partial^2 p}{\partial z^2} + N^2 \nabla_h^2 p \right]$$

so that comparing with the earlier equation, we have, finally,

$$\Omega = (\rho_0 f N^2) q \qquad (11.24)$$

Thus, the conserved quantity in the wave equation for the pressure is a simple multiple of the potential vorticity. This has very important consequences:

1. Since $\partial q / \partial t = 0$, the initial data that gives the pv is sufficient to determine Ω, and it remains unaltered throughout the motion;
2. The oscillating part of the wave field *has no potential vorticity*. This also follows from the conservation of potential vorticity, since if the motion is periodic, the conservation equation $\partial q / \partial t = 0$ becomes simply $\omega q = 0$. If the frequency is not zero, the potential vorticity must be zero.

Therefore, the pressure and velocity field may be divided into two parts. There is a wave part that carries no potential vorticity and a steady part, which is a steady particular solution of the *p* Eq. 11.3.

Let's write the total solution for *p*, *b* and the velocity:

$$p = p_g(x, y, z) + p_w(x, y, z, t)$$
$$b = b_g(x, y, z) + b_w(x, y, z, t) \qquad (11.25)$$
$$\vec{u} = \vec{u}_g(x, y, z) + \vec{u}_w(x, y, z, t)$$

where the g subscripted variables are independent of time and the w subscripts refer to the wave-like part of the motion.

For the steady, linear part of the motion, the balances are

$$f v_g = \frac{1}{\rho_0} \frac{\partial p_g}{\partial x} \qquad (11.26a)$$

$$f u_g = -\frac{1}{\rho_0} \frac{\partial p_g}{\partial y} \qquad (11.26b)$$

$$w_g = 0 \qquad (11.26c)$$

$$b_g = -\frac{\partial p_g}{\partial z} \qquad (11.26d)$$

that is, for the steady part of the solution, the horizontal velocity is in geostrophic and hydrostatic balance, the vertical velocity is zero and most importantly,

$$q = \zeta_g - f \frac{\partial b_g}{\partial z} / N^2 = \frac{1}{\rho_0 f} \nabla_h^2 p_g + \frac{f}{N^2} \frac{\partial^2 p_g}{\partial z^2} \qquad (11.27)$$

(Note, it is because the geostrophic w is zero that the wave equation for w does not contain the extra time derivative that the pressure equation does, because there is no nontrivial steady solution for w).

Thus, if the potential vorticity is determined by the initial conditions, the above elliptic equation for the geostrophic pressure completely determines the steady part of the solution, since once the geostrophic pressure is determined, both the steady density perturbation and the geostrophic horizontal velocities can be calculated from the geostrophic pressure. Note that the vertical velocity in the steady geostrophic solution is zero as a consequence of the steady form of the adiabatic equation. Thus, the steady geostrophic part of the solution can be determined *independently of the wave part*, in terms of the initial value of the potential vorticity.

Since the wave field carries no pv, the wave part of the pressure field is determined from the homogeneous part of Eq. 11.3; thus,

$$\frac{\partial^2}{\partial t^2}\nabla^2 p_w + f^2 \frac{\partial^2 p_w}{\partial z^2} + N^2 \nabla_h^2 p_w = 0 \tag{11.28}$$

The initial conditions on the wave pressure must satisfy that part of the initial pressure field, which contains no pv. Thus, if the total initial perturbation pressure is $p'_{total}(x,y,z,0)$,

$$p_w(x,y,z,0) = p'_{total}(x,y,z,0) - p_g(x,y,z) \tag{11.29}$$

We will see later how to exploit the potential vorticity conservation for the initial value problem, but now let's return to the wave problem. Let

$$w = W_0 e^{i(kx+ly+mz-\omega t)} \tag{11.30}$$

substitution in the wave equation yields the dispersion relation,

$$\omega^2 = f^2 \frac{m^2}{k^2+l^2+m^2} + N^2 \frac{k^2+l^2}{k^2+l^2+m^2} \tag{11.31}$$

Note that if m were zero, the frequency would be N, while if k and l were zero, the frequency would be f. Let K_h be the magnitude of the horizontal component of the wave vector , i.e., $K_h = \sqrt{k^2+l^2}$, such that (see Fig. 11.2)

$$\sin\theta = m/K \tag{11.32a}$$

$$\cos\theta = K_h/K \tag{11.32b}$$

$$K = \left[k^2+l^2+m^2\right]^{1/2} \tag{11.32c}$$

so that

$$\omega^2 = f^2 \sin^2\theta + N^2 \cos^2\theta \tag{11.33}$$

Since

$$N^2 - \omega^2 = N^2 - f^2\sin^2\theta - N^2\cos^2\theta$$

$$= (N^2 - f^2)\sin^2\theta \tag{11.34}$$

$$\omega^2 - f^2 = (N^2 - f^2)\cos^2\theta$$

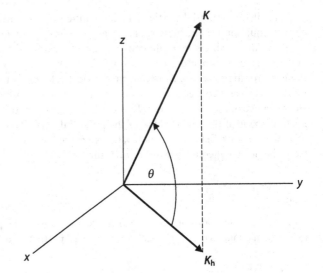

Fig. 11.2. The *three-dimensional wave vector*, whose orientation determines the frequency of the inertial-gravity wave

and since almost everywhere in the ocean $N \gg f$, it follows that the frequency of the plane wave will satisfy $f \le \omega \le N$, which explains the rapid fall off of the spectrum of observed internal waves at N. The observed peak at f is related to the geometrical properties of the forcing. Usually, the horizontal scale is much greater than the vertical scale. In this case, a convenient way to write the frequency relation is

$$\omega^2 = f^2 + (N^2 - f^2)\frac{K_h^2}{K_h^2 + m^2} \approx f^2 + N^2\frac{K_h^2}{m^2}$$

so if $K_h \ll m$, the excited frequencies will be close to f.

With the dispersion relation for the frequency, the student should check that the pv in the wave is zero by calculating both the relative vorticity and b, as well as forming q.

Now let us orient the coordinate system so that the wave vector lies in the x-z-plane. This makes the y-wave number zero. Note *that velocity in the y-direction will be different from zero, since*

$$\frac{\partial v_w}{\partial t} = -f u_w$$

if l is zero.

Then

$$\omega^2 = f^2\frac{m^2}{m^2 + k^2} + N^2\frac{k^2}{m^2 + k^2} \tag{11.35a}$$

$$\omega\frac{\partial \omega}{\partial k} = N^2 k m^2 / K^4 - f^2 k m^2 / K^4 \tag{11.35b}$$

so that

$$\frac{\partial \omega}{\partial k} = \left[N^2 - f^2 \right] \frac{m^2 k}{K^4} \tag{11.36a}$$

and similarly,

$$\frac{\partial \omega}{\partial m} = -\left[N^2 - f^2 \right] \frac{k^2 m}{K^4} \tag{11.36b}$$

Thus the components of the group velocity are

$$(c_{gx}, c_{gz}) = \frac{(N^2 - f^2)}{\omega K^4} mk(m, -k) \tag{11.37}$$

Once again, the vertical component of the group velocity is opposite to the vertical phase speed

$$\frac{\omega}{m} c_{gz} < 0$$

and the group velocity is perpendicular to the wave vector (because, again, with an incompressible fluid, a three-dimensional plane wave has its fluid velocity orthogonal to the wave vector, and the group velocity will be in the direction of the energy flux vector $\overline{p\vec{u}}$).

In the limit where $k \ll m$, the dispersion relation is

$$\omega^2 = f^2 + N^2 \frac{k^2}{m^2} \tag{11.38}$$

and waves with such frequencies are called *Poincaré waves*.

Now that w or the wave pressure is determined, it is easy to calculate each of the other velocity components and the density perturbation so that the energy in the plane wave may be found. The student should check *whether* equipartition between kinetic and potential energy obtains for the internal gravity waves in the presence of rotation.

If

$$w = W_0 \cos(kx + mz - \omega t)$$

it follows that

$$\frac{p}{\rho_0} = -\frac{(N^2 - f^2)}{m\omega} W_0 \cos(kx + mz - \omega t) \tag{11.39a}$$

$$u = -\frac{m}{k} w = -\frac{m}{k} W_0 \cos(kx + mz - \omega t) \tag{11.39b}$$

$$\frac{\partial v}{\partial t} = -fu \Rightarrow v = \frac{f}{\omega} \frac{m}{k} W_0 \sin(kx + mz - \omega t) \tag{11.39c}$$

$$b = -(N^2 / \omega) W_0 \sin(kx + mz - \omega t) \tag{11.39d}$$

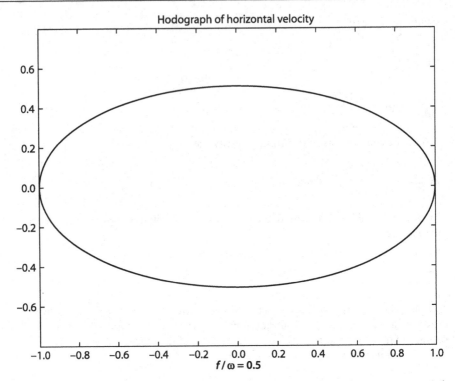

Fig. 11.3. The ellipse traced out by the position of the horizontal velocity vector during one wave period

Note that the relations for u and v imply that the horizontal velocity vector will rotate in the horizontal plane clockwise with time for $\omega > 0$ and $m > 0$, i.e., for downward energy propagation. Indeed, the horizontal velocity vector traces out an ellipse (Fig. 11.3)

$$u^2 + \frac{v^2}{f^2/\omega^2} = W_0^2 \frac{m^2}{k^2} \tag{11.40}$$

whose major axis is along the x-axis and whose minor axis, along the y-axis, is smaller by a factor f/ω.

Finally, note that if the fluid is contained between two lateral boundaries a distance D apart, the equation for the normal modes in that region will be (with the rigid lid approximation)

$$w = W(z)e^{i(kx+ly-\omega t)} \tag{11.41a}$$

$$\frac{d^2 W}{dz^2} + K_h^2 \frac{(N^2 - \omega^2)}{(\omega^2 - f^2)} W = 0 \tag{11.41b}$$

$$W = 0, \quad z = 0, -D \tag{11.41c}$$

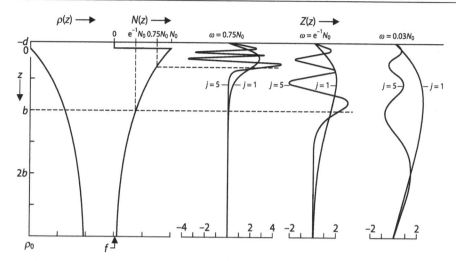

Fig. 11.4. Normal modes for three frequencies. The density distribution and the distribution of $N(z)$ are shown in the first two panels on the left (redrawn from Garrett and Munk 1976)

The mode shapes will depend on the frequency, and the discussion of the domains of oscillation and evanescence in z are similar to the nonrotating case. Figure 11.4, taken from the previously referenced article by Garrett and Munk (1976), shows some examples.

Large-Scale Hydrostatic Motions

For many motions in both the ocean and the atmosphere, the horizontal scale far exceeds the vertical scale of the *motion*. For example, motions in the ocean occurring in the thermocline will have a vertical scale of a kilometer or less, while the horizontal scales might be of the order of hundreds of kilometers. Motions in the ocean induced by traveling meteorological systems will have such large scales. If the *motion* has such disparate scales in the vertical and horizontal, we can expect important influences on the dynamics. First of all, we would expect that the vertical velocity will be small compared with the horizontal velocity, since the motion consists of *nearly* flat trajectories. That in turn could mean that the vertical acceleration is small. Such dynamical consequences often allow simplifications to our treatment of the physics, and we are always looking for such simplifications so that we can make progress with *more difficult* problems; not just make life easier for ourselves.

We need to define clearly what "small" means in a dynamical context. As an example, let's review the results of the plane internal gravity wave with rotation. As we saw in the last lecture, if

$$w = W_0 \cos(kx + mz - \omega t) \tag{12.1}$$

then the pressure and horizontal velocity is

$$\frac{p}{\rho_0} = -\frac{(N^2 - \omega^2)}{m\omega} W_0 \cos(kx + mz - \omega t) \tag{12.2a}$$

$$u = -\frac{m}{k} W_0 \cos(kx + mz - \omega t) \tag{12.2b}$$

Therefore, the ratio of the vertical acceleration to the vertical pressure gradient is

$$\frac{\rho_0 w_t}{p_z} = O\left(\frac{\omega^2}{N^2 - \omega^2}\right) \quad \text{while} \quad \omega^2 = N^2 \frac{k^2}{K^2} + f^2 \frac{m^2}{K^2} \quad \text{or} \tag{12.3}$$

$$\frac{\rho_0 w_t}{p_z} = O\left(\frac{f^2 + (N^2 - f^2)k^2/K^2}{(N^2 - f^2)m^2/K^2}\right)$$

$$= \frac{k^2}{m^2} + \frac{f^2}{N^2} K^2/m^2 \tag{12.4}$$

Now consider the case where the horizontal scale is much larger than the vertical scale of the *motion*. This implies that $k \ll m$, and so $K^2/m^2 \approx 1$. Since k/m is small and in oceanographic settings f/N is small, we see that the vertical acceleration is small compared to the vertical pressure gradient. Later we will see more directly how this comes about by scaling the equations of motion, but here we can see it from the solution of our problem. This implies that to

$$O\left[\left(\frac{k}{m}\right)^2, \frac{f^2}{N^2}\right]_{max}$$

the vertical acceleration in the vertical momentum equation can be ignored compared with the vertical pressure gradient so that to this order, that equation is replaced by the *hydrostatic approximation*

$$\frac{\partial p}{\partial z} = -\rho g \tag{12.5}$$

for the perturbation as well as the mean resting state. That is, the motion has such a weak vertical acceleration that although the fluid is in motion, the pressure can be calculated from the hydrostatic equation *as if the fluid were at rest*. This is characteristic of motions whose horizontal length scales for the *motion* are large compared to the vertical scales of the *motion*.

However, note that although w is small with respect to u,

$$\frac{\partial w}{\partial z} = O(mW_0) \tag{12.6a}$$

$$\frac{\partial u}{\partial x} = O(W_0 k(m/k)) = O\left(\frac{\partial w}{\partial z}\right) \tag{12.6b}$$

so that w *cannot be neglected in the continuity equation*. The *vertical velocity* is small with respect to the *horizontal velocity*, but the fast derivative in z in comparison with x compensates.

We will examine such hydrostatic motions and waves starting with a simple homogeneous model.

Potential Vorticity: Layer Model

Consider a layer of inviscid fluid with, to begin with, a flat bottom and a uniform density ρ. The fluid is rotating with a constant angular velocity $\Omega - f/2$, whose axis is opposite to the gravitational force (see Fig. 12.1).

This is meant to be a model for a small segment of the ocean, whose lateral scale, while being much greater than the depth, is small enough so that the dynamical effects of the Earth's sphericity can be ignored. We therefore use Cartesian coordinates. Further, it is convenient to define $P = p/\rho$ in terms of which the linearized equations of motion are, when the hydrostatic approximation is used,

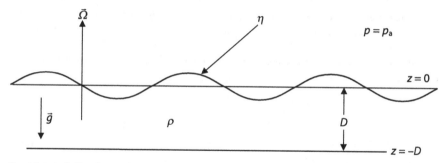

Fig. 12.1. A shallow layer of fluid rotating and of constant density

$$\frac{\partial u}{\partial t} - fv = -\frac{\partial P}{\partial x} \tag{12.7a}$$

$$\frac{\partial v}{\partial t} + fu = -\frac{\partial P}{\partial y} \tag{12.7b}$$

$$0 = -g - \frac{\partial P}{\partial z} \tag{12.7c}$$

$$\frac{\partial u}{\partial x} + \frac{\partial v}{\partial y} + \frac{\partial w}{\partial z} = 0 \tag{12.7d}$$

From the hydrostatic equation, it follows that

$$P = g(\eta - z) + P_a(x, y, t) \tag{12.8}$$

where η is the free surface height and P_a is the atmospheric pressure field at the free surface. Using the result of the calculation of the pressure, the horizontal momentum equations become (subscripts for derivatives)

$$u_t - fv = -g\eta_x - P_{ax} \tag{12.9a}$$

$$v_t + fu = -g\eta_y - P_{ay} \tag{12.9b}$$

Since the forcing terms on the right-hand side of the momentum equations are independent of z, it is consistent to look for solutions for u and v that are also independent of z. This allows us to integrate the continuity equation immediately to obtain

$$D(u_x + v_y) + w(h) - w(-D) = 0$$

Since w is zero at the bottom ($z = -D$) and is equal to $\partial\eta / \partial t$ at the free surface, the equation for mass conservation is simply

$$\frac{\partial\eta}{\partial t} + D(u_x + v_y) = 0 \tag{12.10}$$

The horizontal convergence of velocity times D yields the horizontal convergence of volume. This must be compensated for by an increase in the free surface elevation η.

The elimination of the pressure from the horizontal momentum equations leads to the vorticity equation

$$\zeta = v_x - u_y \tag{12.11a}$$

$$\frac{\partial \zeta}{\partial t} = -f(u_x + v_y) = \frac{f}{D}\frac{\partial \eta}{\partial t} \tag{12.11b}$$

or

$$\frac{\partial}{\partial t}\left[\zeta - \frac{f}{D}\eta\right] = 0 \tag{12.11c}$$

which is the statement of *potential vorticity conservation* for the linear, single layer model when f is constant. We define the pv as

$$q = \zeta - \frac{f}{D}\eta \tag{12.12}$$

To obtain a wave equation for disturbance, we start by taking the divergence of the horizontal momentum equations:

$$\frac{\partial}{\partial t}\left[u_x + v_y\right] - f\zeta = -g\nabla^2\eta - \nabla^2 P_a \tag{12.13}$$

which with the continuity equation yields

$$\frac{1}{D}\frac{\partial^2\eta}{\partial t^2} - f\zeta = -g\nabla^2\eta - \nabla^2 P_a \tag{12.14}$$

The vorticity can be eliminated from this equation with the aid of the equation relating the vorticity to the potential vorticity so that

$$\nabla^2\eta - \frac{1}{c_0^2}\frac{\partial^2\eta}{\partial t^2} - \frac{f^2}{c_0^2}\eta = \nabla^2 P_a/g - \frac{f}{g}q \tag{12.15}$$

where $c_0 = \sqrt{gD}$ is the gravity wave speed for long waves in a *nonrotating fluid*.

Note that the potential vorticity is, by the conservation statement, *independent of time*. Thus, once again we can separate the solution for η into a steady part (which will be in geostrophic balance) and an unsteady part associated with the waves and which will carry no potential vorticity. The geostrophic part of the field will absorb the consequences of the initial distribution of potential vorticity, while the remainder of the initial conditions, the part containing no pv, will radiate away as gravity waves. Again, the steady part can be calculated independently of the wave part so that the final steady state after the wave has radiated away can be calculated independently of the time-dependent wave problem.[1] Historically, this has given rise to a set of interesting *adjustment problems*, starting with the classical paper of Rossby (1938) (see also Gill 1982). In these investigations, the question is asked: suppose we start with an initial distribution of velocity and free surface elevation *not* in geostrophic balance. How does the fluid adjust to eventual geostrophic balance, and what is the final geostrophically balanced state? The first part of the question requires the solution of the wave radiation problem (not easy), while the second part, the ultimate geostrophic state, is *very* easy because of the conservation of pv. We will give a classical example here to see how this works. It should be clear that the process is of general application when pv is conserved.

Rossby Adjustment Problem

Consider a layer of fluid in which at time $t = 0$, a slab of fluid occupying the range between a and $-a$ set into motion with a uniform velocity U along the x-axis, and at the same initial instant the free surface elevation is zero (Fig. 12.2). Suppose the atmospheric pressure forcing is zero. We would expect, somehow, that eventually the free surface will deform, producing a pressure gradient in the y-direction to balance at least part of the initial x-velocity. The question is of the original motion: how much ends up in steady geostrophic balance and how much of the original energy is radiated away in the form of gravity waves? Thus,

$$v = \eta = 0, \quad t = 0 \tag{12.16a}$$

$$u = \begin{cases} U & |y| < a \\ 0 & |y| > a \end{cases} \tag{12.16b}$$

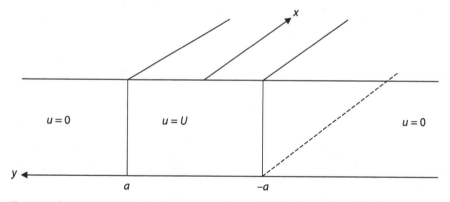

Fig. 12.2. The initial condition in a homogeneous layer of fluid before geostrophic adjustment

---

[1] We are assuming that the atmospheric pressure forcing has no steady part. Otherwise it is easy to show that the response to such forcing is just an inverted barometer response in which the velocity is zero and $\eta = -P_a/g$, a rather dull solution from the point of view of wave dynamics.

The potential vorticity, q, at $t = 0$ can be easily calculated. The initial free surface height is zero, so the only contribution to the potential vorticity comes from the relative vorticity:

$$\zeta = -\frac{\partial U}{\partial y} = -U[\delta(y+a) - \delta(y-a)] \tag{12.17}$$

where $\delta(x)$ is the Dirac delta function. It is zero except where its argument is zero, where its value is infinite and has the property that its integral over the origin of its argument is one. It is the derivative of the step function $H(x)$. Since the original zonal velocity can be written as the sum of two step functions

$$u = U[H(y+a) - H(y-a)]$$

(see Fig. 12.3) the result for the vorticity follows directly.

The potential vorticity is thus

$$q = \left(\zeta - \frac{f}{D}\eta\right)_{t=0} = -U(\delta(y+a) - \delta(y-a)) \tag{12.18}$$

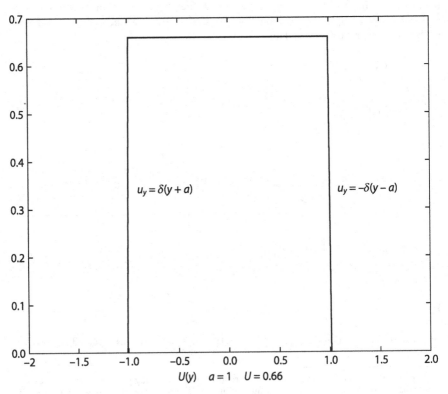

Fig. 12.3. The initial velocity as a function of y showing the two delta functions in the vorticity

In the steady state in which the final adjusted state is to be calculated,

$$\nabla^2\eta - \frac{f^2}{c_0^2}\eta = -\frac{Uf}{g}\left[\delta(y+a) - \delta(y-a)\right] \tag{12.19}$$

Note that the problem is forced entirely by the distribution of pv. Since we expect u and η to be functions only of y in the final state, the equation can be simplified to

$$\eta_{yy} - \frac{f^2}{c_0^2}\eta = -\frac{Uf}{g}\left[\delta(y+a) - \delta(y-a)\right] \tag{12.20}$$

which must be solved on the infinite y-interval between plus and minus infinity. Since the delta functions are zero except at the zeros of their arguments, the right-hand side of the above equation is zero except at the two points $\pm a$. Note that what we are really doing is finding the Greens function response for η to two point sources of pv at the two points $\pm a$.

The velocity field u is initially even about the origin $y = 0$, and there is nothing in the linear problem above that will break that symmetry for u. If u is an even function of y, η, whose derivative with respect to y yields the geostrophic u, must be an odd function of y.

Thus the solution can be written

$$\eta = \begin{cases} A\sinh(y/\lambda) & |y| \leq a \\ Ae^{-(y-a)/\lambda}\sinh(a/\lambda) & y \geq a \\ -Ae^{(y+a)/\lambda}\sinh(a/\lambda) & y \leq -a \end{cases} \tag{12.21}$$

where λ is the *deformation radius* defined by

$$\lambda = \frac{c_0}{f} \tag{12.22}$$

The deformation radius is an intrinsic length scale and measures the tendency for gravity to smooth disturbances out horizontally against the tendency for rotation to link the fluid together vertically along the rotation axis. If the fluid were stratified, instead of c_0, the appropriate speed for defining the deformation radius would be the internal gravity wave speed for a particular vertical mode. Hence, for a stratified fluid there will in general be an *infinite number* of deformation radii. In the present case, we have only one for the homogeneous layer. Like many other fundamental quantities in GFD, this one is named after Rossby and is often called the *Rossby deformation radius*. We will shortly see why the word deformation is used.

We have used the anticipation of antisymmetry for η to write the solution in terms of a single unknown constant A. We have also chosen the solution so that the free surface elevation is continuous at the point $\pm a$. Otherwise, since u is proportional to the y-derivative of η, we would generate infinite velocities at those points.

To determine the constant A, we return to the differential equation for η and integrate it over a small neighborhood around the point $y = a$, i.e., from $y = a - \varepsilon$ to $y = a + \varepsilon$. We will then let $\varepsilon \longrightarrow 0$. Carrying out the integration and remembering that the integral of the delta function is unity when the interval includes the zero of its argument, we obtain

$$\eta_y(a+\varepsilon) - \eta_y(a-\varepsilon) = Uf/g \tag{12.23}$$

In the limit where $\varepsilon \longrightarrow 0$, this yields, using the limits of the solutions on each side of the point $y = a$,

$$-\frac{A}{\lambda}\sinh(a/\lambda) - \frac{A}{\lambda}\cosh(a/\lambda) = Uf/g$$
$$\Rightarrow A = -\frac{\lambda U f}{g}e^{-a/\lambda} \tag{12.24}$$

This completes the solution. Collecting our results, we have

$$\eta/D = \frac{U}{c_0}\begin{cases} -e^{-a/\lambda}\sinh(y/\lambda) & |y| \le a \\ -e^{-y/\lambda}\sinh(a/\lambda) & y \ge a \\ e^{y/\lambda}\sinh(a/\lambda) & y \le -a \end{cases} \tag{12.25}$$

from which the geostrophic zonal velocity of the final state can be calculated from

$$u = -\frac{g}{f}\eta_y$$

$$u/U = \begin{cases} e^{-a/\lambda}\cosh(y/\lambda) & |y| \le a \\ -e^{-y/\lambda}\sinh(a/\lambda) & y \ge a \\ -e^{y/\lambda}\sinh(a/\lambda) & y \le -a \end{cases} \tag{12.26}$$

Note that u is *not* continuous as the point $+a$ and $-a$; the jumps in the velocity of the initial conditions persist to the final steady state that is forced by delta function sources of potential vorticity at $\pm a$, which give rise to "kinks" in the free surface elevation at those points where the slope of η is discontinuous. Note that v is zero in the steady state, although it is certainly not zero in the waves whose radiation is essential to reach the steady state.

The solution for the adjusted steady state has some curious and nonintuitive properties.

Figure 12.4 shows the solution for η and u for the case where $a = 1$ and λ is 10, i.e., when the deformation radius is large compared to the geometrical scale of the flow.

Note that in the figures the zonal velocity, whose profile is shown coming out of the paper, is positive.

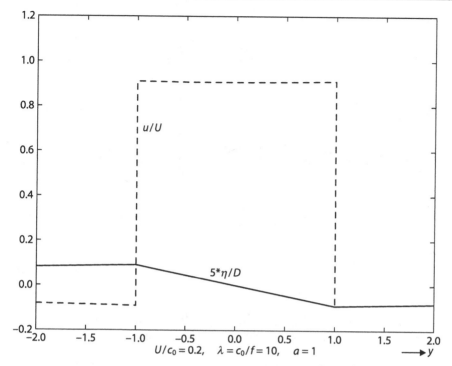

Fig. 12.4. The free surface height (*solid*) and the *x*-velocity (*dashed*) for the case where the deformation radius is ten times the current half width *a*

The velocity profile has not changed much; it is still nearly the sum of step functions of the initial data, but the free surface elevation that was initially zero has changed. The fluid, under the influence of the Coriolis force, has slid to the right of the direction of flow to set up a pressure gradient with high pressure to the right of the current and low pressure to the left of the current (looking downstream). This final adjusted state seems intuitively attractive, and indeed is often the example used for illustrative purposes and close to the one Rossby originally used (note: the free surface height has been multiplied by 5 for clarity).

The results become a good deal stranger when the deformation radius is as small or smaller than the geometrical scale, i.e., when $\lambda < a$. For example, when λ is equal to a, we get the situation shown in Fig. 12.5.

Note that now the reduction of the zonal velocity in the center of the region is much more evident. The free surface is tilting to support the flow geostrophically, but note the reverse flow in the region beyond $\pm a$. Also note that the characteristic decay scale for the deformed free surface is just the deformation radius, hence the name. The really fundamental role of the potential vorticity is particularly evident when the deformation radius is small with respect to the scale a. Figure 12.6 shows the case when λ is 0.1a.

In this limit, the final flow consists of two vortex sheets limited to regions of the order of the deformation radius around each edge where the delta functions of the pv are maintained. The free surface elevation is symmetric about each delta function, and the two are nearly nonoverlapping. The structure is very distant from the original pic-

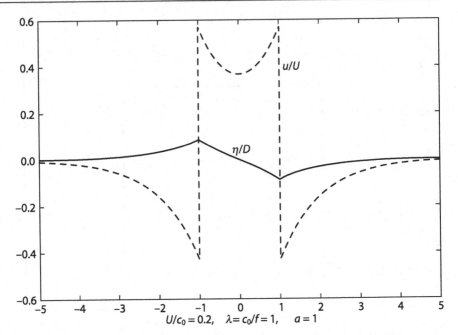

Fig. 12.5. As in Fig. 12.4, except that the deformation radius equals the current half width

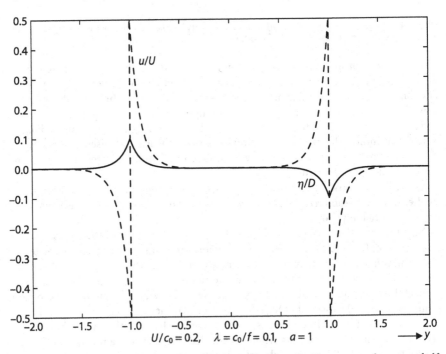

Fig. 12.6. As in Fig. 12.4 except that the deformation radius is small with respect to the current half width, i.e., 0.1

ture of the first figure and emphasizes that the resulting adjusted geostrophic flow is determined by the distribution of pv, *not* by the original distribution of momentum.

Given the remarkable difference between the final states and the initial state, it is important to realize that it is the waves, which are not described, that radiate away that part of the initial condition that will not move to geostrophic balance. One measure of the amount of radiation is the difference between the initial energy and the final energy in the adjusted state.

Energy

By forming the product of the horizontal momentum equations with each velocity component and the product of the continuity equation with the free surface height, it is easy to show that the equation for the energy per unit horizontal area is (in the absence of atmospheric pressure forcing)

$$\frac{\partial}{\partial t}\left[D\left(\frac{u^2+v^2}{2} \right) + \frac{g\eta^2}{2} \right] + \nabla \cdot (g\eta\bar{u})D = 0 \tag{12.27}$$

Note that the kinetic energy involves only the horizontal velocity. Consistent with the hydrostatic approximation, the vertical velocity is too small to contribute. The energy flux term is the horizontal velocity times the pressure, which in this case is given hydrostatically by the free surface elevation.

In the adjustment problem just discussed, the initial energy is all kinetic energy, since the initial free surface elevation is zero. That initial total energy is

$$E_{\text{initial}} = \frac{DU^2}{2} \times 2a = DU^2 a \tag{12.28}$$

We could compute the final energy by using our results for u and η and integrating over the whole y-interval. There is, though, an easier indirect way to do the calculation. Starting from the equation relating η and the potential vorticity,

$$\frac{g}{f}\nabla^2\eta - \frac{f}{D}\eta = q \tag{12.29}$$

Multiplication by $gD\eta/f$ yields

$$D\frac{g^2}{f^2}\nabla \cdot (\eta\nabla\eta) - D\frac{g^2}{f^2}(\nabla\eta)^2 - g\eta^2 = q\eta\frac{gD}{f} \tag{12.30}$$

If the disturbance vanishes at infinity, the divergence term, the first term on the left-hand side of the equation, will have zero integral over the whole domain. Recognizing that the kinetic energy of the geostrophic velocities is given by the second term (divided by 2), we finally obtain for the geostrophically balanced state

$$\int E_{\text{geos.}} dA = -\frac{gD}{2f} \int \eta q \, dA$$

where the integral is over the whole domain. In the case just considered, q is the sum of two delta functions, so

$$\int_{-\infty}^{\infty} E_{geos.} dy = -\frac{c_0^2}{2f} \int_{-\infty}^{\infty} \eta U \left[\delta(y-a) - \delta(y+a) \right] dy$$

$$= -\frac{c_0^2 U}{2f} \left(\eta(a) - \eta(-a) \right) \qquad (12.31)$$

$$= \frac{U^2 D}{2} \lambda \left(1 - e^{-2a/\lambda} \right)$$

The ratio of the final energy to the initial energy will give us a measure of how much is retained in the geostrophic state and how much is radiated away by the gravity waves. That ratio is

$$\frac{E_{geos.}}{E_{initial}} = \frac{1 - e^{-2a/\lambda}}{2a/\lambda} \qquad (12.32)$$

so that the ratio is a function only of the parameter $a/\lambda = fa/c_0$, i.e., the ratio of the width of the current to the deformation radius. The energy ratio as a function of that parameter is shown in Fig. 12.7.

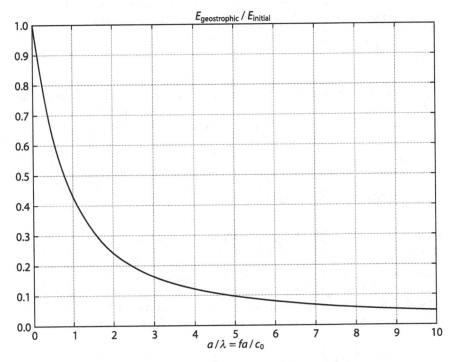

Fig. 12.7. The ratio of the final geostrophic energy to the initial energy

The energy ratio is unity when the deformation radius is large, i.e., when the rotation is negligible. That limit is easy to understand, since if there were no rotation, the initial state of no free surface elevation and a uniform flow in the region between $y = +a$ and $-a$ would be an exact steady solution. Nothing would happen, and no energy would be radiated. At the other extreme, when the deformation radius is small with respect to a, the motion is limited to a narrow region around the end points at $\pm a$, and the energy, almost all of which is kinetic, is of the order of $U^2 D^* \lambda$, i.e., it is of the order of λ. As the deformation radius becomes small, the energy retained in the geostrophic motion becomes small, and in this limit, most of the energy of the initial state is radiated away as gravity waves.

The complete problem was studied by Cahn (1945), an associate of Rossby's. Cahn presents the full-time dependent solution showing the evolution to the time independent state. The analysis is more complicated than we have space for, but I encourage you to at least examine the graphical results of his calculation.

Shallow Water Waves in a Rotating Fluid;
Poincaré and Kelvin Waves

We now examine the nature of the waves which serve, among other things, to sculpt the geostrophic final state from an arbitrary initial state. These waves, as we noted earlier, have no potential vorticity, because in the simple models we are examining, the conservation of pv is simply the statement:

$$\frac{\partial q}{\partial t} = 0 \tag{13.1a}$$

$$q = \zeta - \eta \frac{f}{D} \tag{13.1b}$$

Thus, for periodic motion for which the time derivative can be replaced by multiplication by the frequency, q must vanish. The wave part then satisfies the wave equation derived in the last lecture:

$$\nabla^2 \eta - \frac{1}{c_0^2} \frac{\partial^2 \eta}{\partial t^2} - \frac{f^2}{c_0^2} \eta = 0, \quad c_0 = \sqrt{gD} \tag{13.2}$$

If there were no rotation, we would get the classical, nondispersive wave equation. In one dimension that equation would be

$$\eta_{xx} - \eta_{tt}/c_0^2 = 0 \tag{13.3}$$

Its well-known solution is

$$\eta = F(x - c_0 t) + G(x + c_0 t) \tag{13.4}$$

where F and G are arbitrary functions of their arguments. The functions can be determined by initial data. The important thing to note here is that the shape of the disturbance remains fixed with time, and each function translates F to the right and G to the left with the speed c_0. The unchanging shape is a reflection of the fact that for nonrotating shallow water, the phase speed is independent of wave number; the wave is nondispersive, and so no change of shape occurs.

Returning to the case where the rotation is different from zero, we can find plane wave solutions in the x-y-plane of the form

$$\eta = Ae^{i(kx+ly-\omega t)} \tag{13.5a}$$

$$\vec{K} = \hat{x}k + \hat{y}l \tag{13.5b}$$

$$K = \sqrt{k^2 + l^2} \tag{13.5c}$$

which yields for the frequency the dispersion relation:

$$\omega = \pm\left\{f^2 + c_0^2[k^2 + l^2]\right\}^{1/2} \tag{13.6}$$

It is important to note that the horizontal velocity has a nonzero divergence, since

$$\nabla_h \cdot \vec{u} = -\frac{1}{D}\eta_t$$

so the group velocity does not have to be perpendicular to the wave vector. Indeed, we can tell immediately from the dispersion relation that the group velocity will be in the direction of the wave vector.

Note that the frequency has a minimum value of $\pm f$. That is, these waves all have frequencies greater than the Coriolis parameter. If λ is the wavelength, the increase of the frequency of the wave above f will depend on the ratio of the wavelength to the deformation radius, c_0/f. If the wavelength is large compared to the deformation radius, the frequency will be close to f.

We may easily calculate the two components of the group velocity:

$$\frac{\partial \omega}{\partial k} = c_0^2 \frac{k}{\omega} = c_0^2 \frac{k}{\sqrt{f^2 + c_0^2 K^2}} \tag{13.7a}$$

$$\frac{\partial \omega}{\partial l} = c_0^2 \frac{l}{\omega} = c_0^2 \frac{l}{\sqrt{f^2 + c_0^2 K^2}} \tag{13.7b}$$

Thus, the group velocity is in the direction of the wave vector and in the same direction as the phase speed. Note that while the group velocity goes to zero as the wave number goes to zero (large wavelengths), the phase speed becomes infinite in that limit; this is another indication of the physical irrelevance of the phase speed as a messenger of real information.

By eliminating v between the two horizontal momentum equations, one obtains a simple relation between u and η, i.e.,

$$\frac{\partial^2 u}{\partial t^2} + f^2 u = -g\frac{\partial^2 \eta}{\partial x \partial t} - fg\frac{\partial \eta}{\partial y} \tag{13.8}$$

similarly for v,

$$\frac{\partial^2 v}{\partial t^2} + f^2 v = -g\frac{\partial^2 \eta}{\partial t \partial y} + gf\frac{\partial \eta}{\partial x} \tag{13.9}$$

This allows us to solve for u and v in terms of η unless the operator of the left-hand side is null, which will happen for oscillations exactly at the *inertial frequency*, i.e., when $\omega = \pm f$.

For all other frequencies, we have the relations (after aligning the x-axis with the wave vector):

$$\eta = \eta_0 \cos(kx - \omega t) \tag{13.11a}$$

$$u = \frac{\eta_0}{D} \frac{\omega}{k} \cos(kx - \omega t) \tag{13.11b}$$

$$v = \frac{\eta_0}{D} \frac{f}{k} \sin(kx - \omega t) \tag{13.11c}$$

so that again the velocity vector traces out an ellipse in the x-y-plane whose major axis is in the direction of the wave vector, and its minor axis shorter by an amount ω/f is at right angles. For positive frequency and wave number, the velocity vector in the wave moves clockwise as the wave progresses through a period. Note that the fluid velocity is smaller than the phase speed by the (small) parameter η_0/D.

We also note that the maximum group velocity is c_0, and this occurs for the shortest waves. The longest waves have the slowest group velocity. Therefore, were we to do the adjustment problem, we would expect that we would see the short waves speed away from the adjusting current first, and after a long time, a long swell of waves would finally move away from the vicinity of the current. This is exactly what Cahn found as seen in Fig. 13.1 redrawn from his paper.

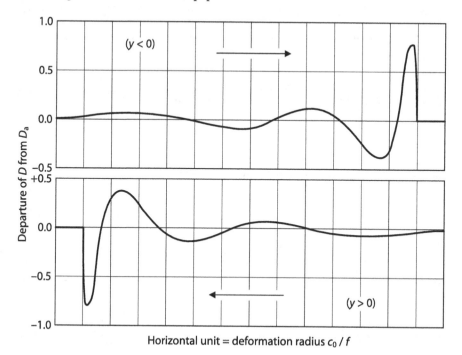

Horizontal unit = deformation radius c_0 / f

Fig. 13.1. The free surface height as a function of time showing at first the passage of the fast short waves and then the longer waves with slower group velocities (after Cahn 1945)

Channel Modes and the Kelvin Wave

Up to now, we have not considered waves in a domain bounded horizontally. Normally, what that does is introduce certain conditions that quantize the horizontal wave number. However, in the case of a rotating fluid, there are some surprises. Consider the wave motion in a channel of width L (Fig. 13.2).

Again, the equation of motion for the wave is

$$\nabla^2\eta - \frac{1}{c_0^2}\frac{\partial^2\eta}{\partial t^2} - \frac{f^2}{c_0^2}\eta = 0, \quad c_0 = \sqrt{gD} \tag{13.12}$$

On the boundaries of the channel, which have been oriented along the x-axis, the y-component of the velocity, v, must vanish. Since

$$\frac{\partial^2 v}{\partial t^2} + f^2 v = -g\frac{\partial^2\eta}{\partial t\partial y} + gf\frac{\partial\eta}{\partial x} \tag{13.13}$$

the boundary condition becomes

$$0 = -g\frac{\partial^2\eta}{\partial t\partial y} + gf\frac{\partial\eta}{\partial x} \tag{13.14a}$$

$$y = 0, L \tag{13.14b}$$

The domain is infinite in the x-direction, and so we can look for wave modes of the form

$$\eta = \overline{\eta}(y)e^{(ikx-\omega t)} \tag{13.15}$$

so that $\overline{\eta}(y)$ satisfies the ordinary differential equation:

$$\frac{d^2\overline{\eta}}{dy^2} + \left\{\frac{\omega^2 - f^2}{c_0^2} - k^2\right\}\overline{\eta} = 0 \tag{13.16}$$

this is subject to the boundary conditions;

$$\frac{d\overline{\eta}}{dy} + k\frac{f}{\omega}\overline{\eta} = 0 \tag{13.17a}$$

$$y = 0, L \tag{13.17b}$$

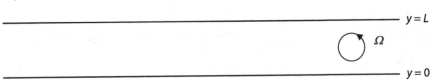

$$y = L$$

$$\Omega$$

$$y = 0$$

Fig. 13.2. The channel of width L in plan view in which gravity modes occur

Note that if there were no rotation, $f = 0$, a possible solution would be $\bar{\eta}$ independent of y with $\omega = \pm k c_0$. This would be the lowest cross stream mode and the solution with the lowest frequency. Higher modes of the form $\cos j\pi y / L$ would also be possible with frequencies

$$\omega = \pm c_0 \sqrt{k^2 + j^2\pi^2 / L^2}$$

It will be of interest to examine how the rotation alters this simple structure of the problem (see also Gill 1982 and Pedlosky 1987). The cause of the change will be found in the *mixed* boundary condition at $y = 0$ and L, which mixes the function and its derivative, and which explicitly involves the frequency.

It is useful to define the constant

$$\ell^2 = \frac{\omega^2 - f^2}{c_0^2} - k^2 \tag{13.18}$$

so that ℓ is something like a y-wave number. The solution can then be written

$$\bar{\eta}(y) = A \sin \ell y + B \cos \ell y \tag{13.19}$$

The constants A and B are not both free but must be chosen to satisfy the boundary conditions. Applying the boundary condition on $y = 0$ first yields

$$A\ell + B\frac{kf}{\omega} = 0 \quad \Rightarrow B = -\frac{\omega \ell}{kf} A \tag{13.20}$$

The same boundary condition applied at $y = L$ yields

$$A\ell \cos \ell L - B\ell \sin \ell L + \frac{kf}{\omega}\{A \sin \ell L + B \cos \ell L\} = 0 \tag{13.20}$$

which when combined with the first equation relating A and B yields

$$\sin \ell L \left[\frac{\omega \ell^2}{kf} + \frac{kf}{\omega} \right] = 0 \tag{13.21}$$

When the definition of ℓ is used to evaluate the square bracket in the condition above, we obtain the final *eigenvalue relation* for the modes in the rotating channel, i.e.,

$$\sin \ell L [\omega^2 \ell^2 + k^2 f^2] = \sin \ell L \left[\omega^2 \left(\frac{\omega^2 - f^2}{c_0^2} \right) - k^2 \omega^2 + k^2 f^2 \right]$$

$$= \sin \ell L \left[\frac{\omega^2}{c_0^2} - k^2 \right] (\omega^2 - f^2) = 0 \tag{13.22}$$

There are apparently three possible ways in which this eigenvalue relation, or dispersion relation linking ω and k, can be found.

1. $\omega = \pm f$

2. $\omega = \pm k c_0$

3. $\sin \ell L = 0$

The first of these is immediately suspicious, since as we have noted, when the frequency is exactly equal to f, we can no longer use Eqs. 13.8 and 13.9 that relate the velocities to the free surface elevation. We will have to examine this case very carefully, and in fact, it will turn out that this is a spurious root.

The second possibility does not look much more promising, since it appears to yield the frequency relation for a y-independent mode for a nonrotating fluid. Note that the boundary conditions

$$\frac{d\overline{\eta}}{dy} + k\frac{f}{\omega}\overline{\eta} = 0, \quad y = 0, L \tag{13.23}$$

do not allow a nontrivial y-independent solution if f is not zero.

With all this doubt in mind, let's start with the term that looks like the most ordinary of eigen conditions, namely

$$\sin \ell L = 0 \tag{13.24}$$

The solution of this condition is

$$\ell L = n\pi, \quad n = 1, 2, 3... \tag{13.25}$$

where we note that we have started with $n = 1$. The solution corresponding to $n = 0$ would yield, from the boundary condition at $y = 0$, $B = 0$. But if $\ell = 0$, the remaining term proportional to A would be the sine of a zero argument. Hence, the whole solution becomes trivially zero.

The physical reason why this occurs is related to the relation between η and v. If the free surface height *were* independent of y, we would have

$$v = \frac{gf}{f^2 - \omega^2}\frac{\partial \eta(x,t)}{\partial x} \tag{13.26}$$

which would be nonzero at the boundaries $y = 0, L$, unless η were identically zero everywhere. Hence, the first nontrivial term must be $n = 1$. Using the definition

$$\ell^2 = \frac{\omega^2 - f^2}{c_0^2} - k^2 \tag{13.27}$$

this yields the dispersion relation for ω for each n,

$$\omega_n^2 = f^2 + c_0^2\left[k^2 + n^2\pi^2/L^2\right] \tag{13.28}$$

This is exactly the dispersion relation for the plane Poincaré wave we deduced earlier, except that the y-wave number is quantized in multiples of π/L *with the major exception that the* $n = 0$ *mode is not allowed.* Now, in the unbounded case, there is such a y-independent mode. In addition, when f is zero there is such a mode allowed. What has happened to that lowest mode? Something is missing, since it makes no physical sense that the addition of the smallest rotation of the system can eliminate the lowest mode previously allowed. We have a problem here we must be sure to clear up. For now though, let's go ahead as if we have not noticed this vexing apparent paradox and examine what the modes that are allowed are like.

Using the relation between A and B and choosing A to measure the elevation of the free surface,

$$\eta = \eta_0\left[\cos(n\pi y/L) - \frac{kfL}{\omega n\pi}\sin(n\pi y/L)\right]\cos(kx - \omega t) \tag{13.29}$$

Note that the y-structure depends on the *phase speed* of the mode; that is, in the square bracket the relative importance of the sine term with respect to the cosine term depends on ω/k. To keep the equations uncluttered, the subscript n on the frequency has been suppressed, but the student should recall that for each n, the frequency is given from the dispersion relation above. Since for each k there are two roots for ω differing in sign, it follows that the cross channel structure will differ for waves going to the right and waves going to the left. Using the relations between the velocities and the free surface height

$$v(f^2 - \omega^2) = gf\frac{\partial\eta}{\partial x} - g\frac{\partial^2\eta}{\partial t\partial y} \tag{13.30}$$

one easily finds that

$$v = \frac{-\eta_0}{D}\frac{\left[f^2 + c_0^2 n^2\pi^2/L^2\right]}{\omega n\pi/L}\sin(n\pi y/L)\sin(kx - \omega t) \tag{13.31}$$

The y-component of velocity contains only the sine term, since of course it has to vanish on $y = 0$ and L. Similarly, the velocity in the x-direction can be found and is

$$u = \frac{\eta_0}{D}\left[\frac{c_0^2}{(\omega/k)}\cos(n\pi y/L) - \frac{fL}{n\pi}\sin(n\pi y/L)\right]\cos(kx - \omega t) \tag{13.32}$$

Looking back at the formula for the free surface elevation (Eq. 13.29), we note that η will vanish for those values of y for which

$$\tan(n\pi y/L) = \frac{\omega_n \ell_n}{kf} = \frac{n\pi}{L}\left\{1 + \frac{c_0^2(k^2 + n^2\pi^2/L^2)}{f^2}\right\}^{1/2} \tag{13.33}$$

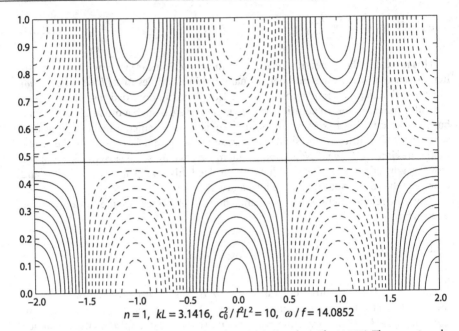

Fig. 13.3. The elevation of the free surface for a low rotation mode, $\omega/f = 14.0852$. The gravest mode corresponding to $n = 1$ is shown

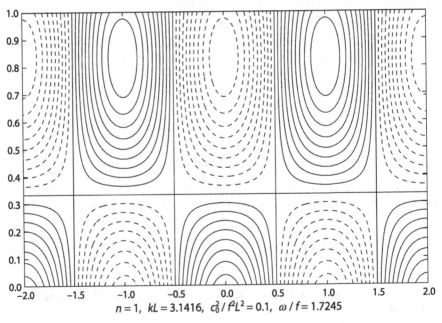

Fig. 13.4. The gravest mode for the case of large rotation. $\omega/f = 1.7245$ and the deformation radius is about 1/3 of the channel width

When $f = 0$, these coincide with the infinities of the tangent function, i.e., at y at odd half multiples of L. So, in the limit as $f \longrightarrow 0$, all these modes coincide with the cosine modes of cross stream wave number $n\pi / L$ for $n > 0$. So, once more we are perplexed to find that even though we have an infinite number of modes, we are missing the lowest mode corresponding to $n = 0$.

Let's first look at the structure of the modes we *have* found. First let's look at the mode for small f or when the deformation radius is much larger than the channel width. For $c_0 / fL = 10$, the contours of the free surface height are shown in Fig. 13.3 for an x-wave number π / L.

Note that for this case, at low rotation the lowest mode has a node at about the halfway point in y in the channel. On the other hand, when the rotation rate is large so that the deformation length is about a third (actually 0.316) of the channel width, the form of the free surface is as shown in Fig. 13.4, and we note that now the zero level is much closer to the lower boundary.

Both of the cases above are for wave patterns propagating to the right. If the frequency is negative so that the pattern propagates to the left with the same speed as the above example, the free surface height instead looks like in Fig. 13.5.

The pattern is essentially the same except that the nodal line in y has shifted towards the boundary at $y = L$ as the pattern propagates to the left.

The student is left to discuss the group velocity in these modes in the x-direction. Note that there is no energy flux in the y-direction, and it is left to the student to explain why that is in terms of individual plane Poincaré waves.

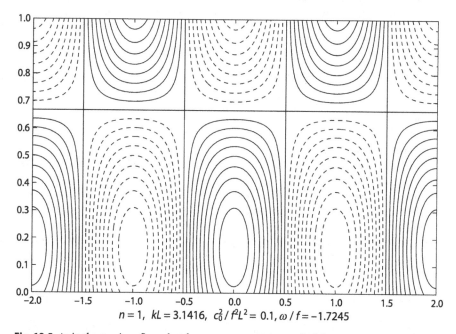

$$n = 1, \ kL = 3.1416, \ c_0^2 / f^2 L^2 = 0.1, \omega / f = -1.7245$$

Fig. 13.5. As in the previous figure but for a wave propagating to the left

The Kelvin Mode

Let's now examine the second possibility as a solution of the eigenvalue condition, namely that

$$\omega = \pm k c_0 \tag{13.34}$$

As we noted, this is a rather unexpected possibility, since it is the dispersion relation for y-independent, nonrotating, long surface gravity waves. Our fluid is rotating, and as we noted above, no solution independent of y is a possible solution in the channel. So, it is of interest to examine the possibility with some care. Using the definition

$$\ell^2 = \frac{\omega^2 - f^2}{c_0^2} - k^2 \tag{13.35}$$

we find that for this case,

$$\ell = \pm i f / c_0 \tag{13.36}$$

so that the cross channel wave number is *purely imaginary*. Let's look at the solution corresponding to the positive imaginary root (it is left to the student to repeat the analysis for the negative root to demonstrate that nothing new is discovered; the negative root only serves to interchange the identity of the two solutions we will shortly find).

Using the relation between A and B from the boundary condition at $y = 0$,

$$A\ell + B\frac{kf}{\omega} = 0, \quad \Rightarrow B = -\frac{\omega \ell}{kf} A \tag{13.37}$$

and writing the sine and cosine in their exponential form, we obtain

$$\overline{\eta}(y) = \eta_0 \left\{ e^{-fy/c_0} [1 + \omega/kc_0] - e^{fy/c_0} [1 - \omega/kc_0] \right\} \tag{13.38}$$

Here we have redefined $A = i\eta_0$.

The solution consists of two parts. The first term exponentially decreases from the lower boundary at $y = 0$. The second term exponentially decreases in the $-y$-direction from boundary happens at $y = L$. The scale for the exponential decrease from either boundary is the deformation radius. Note that this solution can occur only in the presence of lateral boundaries in order to keep the solution finite for all y.

For the solution propagating to the right for which $\omega = kc_0$, the *second term vanishes*, and the total solution restoring the x- and t-dependence is the right moving *Kelvin wave*:

$$\eta = \eta_0 e^{-fy/c_0} \cos(kx - \omega t) \tag{13.39}$$

Note that such a solution would be valid for the region $y > 0$ if only a single wall were present and the fluid were semi-infinite in the $+y$-direction. We obtained the solution using only the boundary condition at $y = 0$, and we must check that it also satisfies the $v = 0$ condition at $y = L$. In fact, let's calculate v for all y in the channel using the relation

$$v(f^2 - \omega^2) = gf\eta_x - g\eta_{yt}$$

$$= -gfk\sin(kx - kc_0 t) + g\frac{f}{c_0}kc_0\sin(kx - kc_0 t) \tag{13.40}$$

$$= 0!$$

The cross channel velocity is *identically zero* for all values of y in the channel, and so of course this satisfies the boundary conditions trivially at $y = 0$ and L. Moreover, calculating u,

$$u(f^2 - \omega^2) = -gf\eta_y - g\eta_{xt}$$

$$= g\frac{f^2}{c_0}\eta - gk^2 c_0\eta \tag{13.41}$$

$$= \frac{g}{c_0}(f^2 - k^2 c_0^2)\eta = \frac{g}{c_0}(f^2 - \omega^2)\eta$$

or

$$u = -\frac{g}{f}\frac{\partial\eta}{\partial y} \tag{13.42}$$

so that the long channel velocity *is in geostrophic balance with the pressure field, although the motion is unsteady and the frequency is not small with respect to* f.

If we choose the other root $\omega = -kc_0$ so that the wave is traveling to the left, the solution consists of the same wave, now a maximum at the boundary at $y = L$ exponentially decreasing in the direction towards the lower boundary at $y = 0$. Again, as you can check, the cross channel velocity is exactly zero, and the long channel velocity is in geostrophic balance. Note that in regions where the free surface elevation and the u-velocity are in phase and if one is positive, so is the other.

Note that as $f \longrightarrow 0$, the mode becomes independent of y and

$$\eta \to \eta_0 \cos(kx - \omega t)$$

which is the "missing" lowest mode of the nonrotating case. That mode in the presence of rotation maintains its character of having no cross channel velocity and does so by introducing a sloping free surface elevation that exactly balances geostrophically the Coriolis acceleration of u. Indeed, it is illuminating to examine the original equations using the a priori condition that v is identically zero, i.e.,

$$fu = -g\eta_y \tag{13.43a}$$

$$u_t = -g\eta_x \tag{13.43b}$$

$$\eta_t = -Du_x \tag{13.43c}$$

Combining the last two of these equations yields

$$\eta_{tt} = c_0^2 \eta_{xx} \tag{13.44}$$

one solution of which is $\eta = F(x - c_0 t, y)$ where F is an arbitrary function. The first equation determines the y-structure of F. Since from that equation $fu_t = -g\eta_{yt}$, we find that from the second equation $-fg\eta_x = -g\eta_{yt} = gc_0\eta_{yx}$, the second equality follows from the $x - c_0 t$ structure of the function F. This yields the differential equation

$$\frac{\partial}{\partial y}\eta_x + \frac{f}{c_0}\eta_x = 0 \tag{13.45}$$

from which the exponential y-structure of the solution follows immediately. An important consequence of this approach is that we see that the form of the Kelvin wave in x is arbitrary. Any function of the argument $x - c_0 t$ is legitimate for the x-t-structure, and as we could see from its dispersion relation, the form in x is unchanging with time as the wave propagates, because the frequency relation is nondispersive, i.e., the frequency is a linear function of k.

While the Poincaré waves have a minimum frequency

$$\omega_{min} = \left[f^2 + c_0^2 \pi^2 / L^2\right]^{1/2} \tag{13.46}$$

the Kelvin wave has no minimum. As $k \longrightarrow 0$, the frequency will go to zero. So if we had a rather narrow channel for which the minimum frequency was quite a bit higher than f, a forcing at or below the inertial frequency would not be possible in the channel for Poincaré waves. The dynamical equations (linear, rotating shallow water) we have been studying are often called the *Laplace Tidal Equations*, because they are exactly those used to discuss the tidal response to solar and lunar forcing. Naturally, one has to include the effects of sphericity, which we have not done, but qualitatively we can see there would be difficulty of the tidal forcing at semi-diurnal or diurnal periods to effective produce a Poincaré wave response in a narrow sea. Instead, the response is more likely to be a Kelvin wave signal propagating around the boundary of the sea.

To get a feeling what that might look like, consider the superposition of two Kelvin waves of equal amplitudes propagating on both boundaries of the channel representing an incoming wave on one boundary balanced by an outgoing wave on the other boundary. We are neglecting the (difficult) problem of the reflection of the Kelvin wave at one of the ends of the channel if it is close, but we will assume it is far enough away to ignore.

The sum of the two Kelvin waves would be

$$\eta = \eta_0 \left[\cos(k[x - c_0 t])e^{-fy/c_0} + \cos(k[x + c_0 t])e^{-(L-y)f/c_0}\right] \tag{13.47}$$

where we have introduced a constant term in the second wave so that each wave has the same maximum amplitude. Note that the second wave is decreasing as y diminishes.

Note that at $y = L/2$, the channel mid-point, the free surface height is

$$\eta = \eta_0 \left[\cos(k[x - c_0 t]) + \cos(k[x + c_0 t]) \right] e^{-fL/2c_0}$$
$$= 2\eta_0 \cos kx \cos kc_0 t \, e^{-fL/2c_0} \qquad (13.48)$$

so that for all t the free surface elevation vanishes and is therefore fixed in time at the points

$$y = L/2, \quad kx = j\pi/2, \quad j = 1,2,3\ldots \qquad (13.49)$$

These fixed points for the elevation are called amphidromic points in tidal theory. Figures 13.6–13.8 show the sum of the two Kelvin waves at several times over a wave period. The asterisks mark the amphidromic points.

Figure 13.6 shows the case where kL is π and the figure is drawn for the time $t = T/4$ where T is the wave period $2\pi/kc_0$. Figure 13.7 shows the free surface elevation some time before when $t = 0.245T$.

Note that the amphidromic points on the zero contour of free surface height have remained stationary as the phase of the disturbance rotates around it. Figure 13.8 shows the situation at the later time $t = 2.55T$.

Again, note that although the phase lines have altered their tilt considerably, the amphidromic points remain stationary.

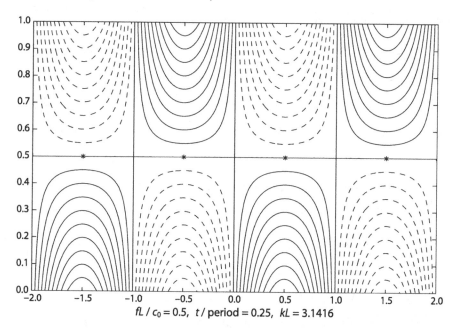

Fig. 13.6. The superposition of a left and right traveling Kelvin wave of equal amplitude in the channel. Note the amphidromic points marked by the asterisks. The pattern is shown at a time where the phase lines in y are vertical

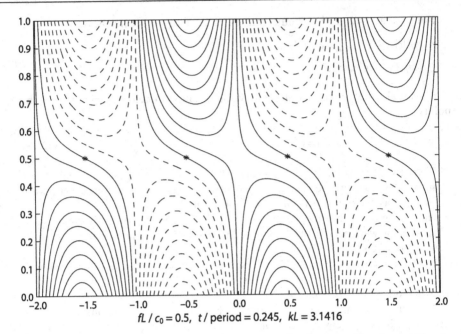

$$fL/c_0 = 0.5, \quad t/\text{period} = 0.245, \quad kL = 3.1416$$

Fig. 13.7. As in the previous figure but at a later time. The phase lines have moved but the amphidromic points are fixed on the phase line

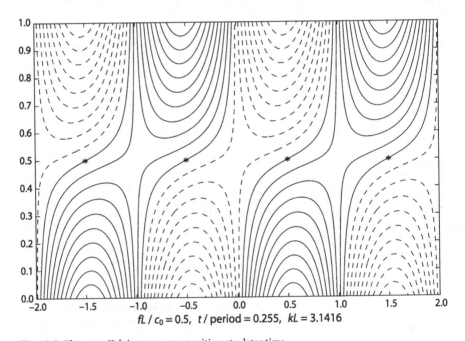

$$fL/c_0 = 0.5, \quad t/\text{period} = 0.255, \quad kL = 3.1416$$

Fig. 13.8. The same Kelvin wave superposition at a later time

We now turn our attention to the third possibility for an eigen solution, namely $\omega = \pm f$, and we noticed that in this case it was not possible to solve directly for the velocity field from the free surface height, since for example

$$\frac{\partial^2 v}{\partial t^2} + f^2 v = -g \frac{\partial^2 \eta}{\partial t \partial y} + gf \frac{\partial \eta}{\partial x} \tag{13.50}$$

the operator on the left-hand side is trivially zero and would give an infinite amplitude for v. We must return to the original equations, i.e.,

$$v_t + fu = -g\eta_y \tag{13.51a}$$

$$u_t - fv = -g\eta_x \tag{13.51b}$$

Let's examine a solution oscillating like e^{-ift}, i.e., with $\omega = f$, and see if it is possible. Then

$$-ifv + fu = -g\overline{\eta}_y \tag{13.52a}$$

$$-ifu - fv = -gik\overline{\eta} \tag{13.52b}$$

Note that the determinant of the coefficients of u and v is zero, but if the second equation is multiplied by i and subtracted from the first, we obtain

$$\frac{d\overline{\eta}}{dy} + k\overline{\eta} = 0 \tag{13.53a}$$

so that

$$\overline{\eta} = \eta_0 e^{-ky} \tag{13.53b}$$

At the same time using *one* of the two momentum equations,

$$u = -iv + \frac{g}{f} k\overline{\eta} \tag{13.54}$$

If this is placed in the equation for mass conservation,

$$\overline{\eta}_t + D(u_x + v_y) = 0 \tag{13.55}$$

we obtain

$$-if\overline{\eta} + ikD(iku + v_y) = 0 \tag{13.56a}$$

$$\Rightarrow \quad -if\overline{\eta} + ikD\left\{ \frac{ifv}{f} - \frac{gk\overline{\eta}}{f} \right\} + Dv_y = 0 \tag{13.56b}$$

$$\Rightarrow \quad v_y - kv = if\frac{\overline{\eta}}{D}\left(1 - \frac{c_0^2 k^2}{f^2}\right) = \frac{if\eta_0}{D}\left(1 - \frac{c_0^2 k^2}{f^2}\right)e^{-ky} \tag{13.56c}$$

which yields the solution for v:

$$v = Ae^{ky} - \frac{if\eta_0}{2Dk}\left(1 - \frac{c_0^2 k^2}{f^2}\right)e^{-ky} \tag{13.57}$$

However, v must vanish on $y = 0$ and L. We can make v vanish on $y = 0$ by the proper choice of A so that

$$v = \frac{if\eta_0}{Dk}\left[1 - \frac{c_0^2 k^2}{f^2}\right]\sinh ky \tag{13.58}$$

The only way v can vanish on $y = L$ is if the coefficient in front of the sinh term vanishes, in which case v is identically zero and the exponential decay rate for η is e^{-fy/c_0}, while the frequency, f, is also kc_0. Thus, there is no possible wave solution with frequency f except at a single wave number at which point the solution is indistinguishable from a Kelvin wave. We conclude that the full solution of the problem consists of an infinite number of Poincaré waves plus the Kelvin wave. The dispersion diagram for the complete problem is shown in Fig. 13.9.

Note that for very large k, all the modes approach the dispersion curve for the Kelvin wave. Note that there are two Kelvin modes, one for each boundary.

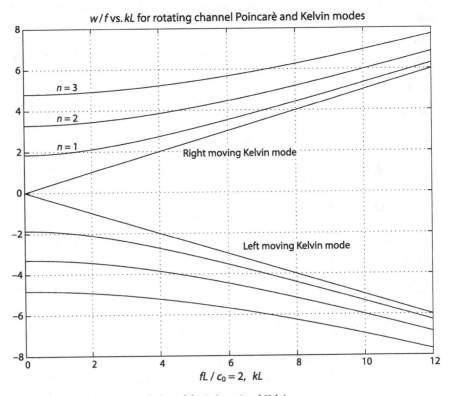

Fig. 13.9. The full dispersion relation of the Poincaré and Kelvin waves

Rossby Waves

When we consider waves of large enough scale, the sphericity of the Earth can no longer be ignored. Rossby was the first to point out that the most significant effect of the Earth's sphericity is that it rendered the Coriolis parameter $f = 2\Omega \sin\theta$, a function of latitude. Since the large scale motions in the ocean are nearly horizontal, the only component of the Coriolis acceleration that really matters is the one involving the horizontal velocities, and therefore only the local vertical component of the Coriolis parameter is dynamically significant. Otherwise, for scales that are large but still sub-planetary, a Cartesian coordinate system can be used to obtain at least a qualitatively correct view of the dynamics. Such an approximation in which the variation of the Coriolis parameter with latitude is treated but in which the geometry is otherwise Cartesian is called the *beta-plane approximation*, and we shall use it without a detailed justification. The student is referred to Pedlosky (1987) for a careful derivation. In this course, we will use the heuristic approach outlined above.

In this way, we take as the governing linear equations of motion

$$u_t - fv = -g\eta_x \tag{14.1a}$$

$$v_t + fu = -g\eta_y \tag{14.1b}$$

where now

$$f = 2\Omega \sin\theta \approx 2\Omega \sin\theta_0 + \frac{2\Omega \cos\theta_0}{R} R(\theta - \theta_0) + \dots$$

$$= f_0 + \beta y, \quad f_0 = 2\Omega \sin\theta_0, \quad \beta = \frac{2\Omega \cos\theta_0}{R} \tag{14.2}$$

where θ_0 is a mid-latitude point about which we have expanded the Coriolis parameter, and R is the Earth's radius. The relation between latitude and the y-variable follows from (see Fig. 14.1)

$$y = R(\theta - \theta_0), \quad \text{and note that} \quad \beta = \partial f / \partial y \tag{14.3}$$

If L is a characteristic magnitude of the north-south scale of the motion, then the change of f compared to its characteristic value is

$$\beta \Delta y / f_0 = O(L/R) \ll 1 \tag{14.4}$$

as the principle parameter restriction for the validity of the beta-plane approximation.

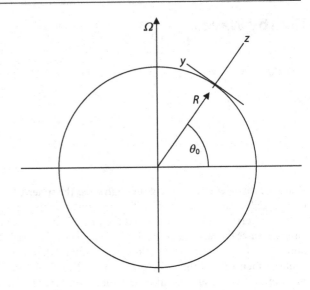

Fig. 14.1.
The tangent β-plane at the
central latitude θ_0

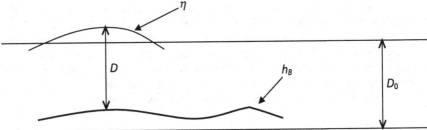

Fig. 14.2. The homogeneous layer of fluid on the β-plane

At the same time, we will let the depth of the fluid *in the absence of motion*, which we have called D, be a function of position as well (Fig. 14.2). Thus,

$$D = D_0 - h_b(x, y) \tag{14.5}$$

Returning to the continuity equation and integrating it over the depth of the fluid, assuming again that since the pressure gradient is independent of depth, we may take the horizontal velocities independent of depth:

$$D(u_x + v_y) + w(\text{top}) - w(\text{bottom}) = 0 \tag{14.6a}$$

$$w(\text{top}) = \eta_t \tag{14.6b}$$

$$w(\text{bottom}) = \vec{u} \cdot \nabla h_b \tag{14.6c}$$

The last condition follows from the kinematic condition that at the bottom, the velocity must be parallel to the bottom so that a horizontal velocity flowing across the gradient of the bottom depth produces a vertical velocity in order that the total velocity is parallel to the bottom. Putting the equations together yields the equation for mass conservation:

$$\eta_t + D(u_x + v_y) + \vec{u} \cdot \nabla D = 0 \tag{14.7a}$$

or

$$\eta_t + \nabla \cdot (D\vec{u}) = 0 \tag{14.7b}$$

(note that $\nabla D = -\nabla h_b$).

Now let's form the vorticity equation by cross-differentiating the momentum equations to eliminate the pressure term. We obtain, remembering that f is a function of y,

$$\zeta_t + f(u_x + v_y) + \beta v = 0 \tag{14.8}$$

and then with the mass conservation equation we can eliminate the horizontal divergence of velocity:

$$\zeta_t + f\left(-\frac{\eta_t}{D} - \frac{\vec{u}}{D} \cdot \nabla D\right) + \beta v = 0 \tag{14.9}$$

or

$$\frac{\partial}{\partial t}\left[\zeta - f\frac{\eta}{D}\right] + \vec{u} \cdot \nabla f - \vec{u} f \cdot \nabla D / D = 0 \tag{14.10a}$$

or equivalently,

$$\frac{\partial}{\partial t}\left[\zeta - f\frac{\eta}{D}\right] + D\vec{u} \cdot \nabla \frac{f}{D} = 0 \tag{14.10b}$$

The first term in this equation is the rate of change of the potential vorticity. The second term is the inner product of the mass flux with the *gradient of the ambient potential vorticity* f / D. Up to now, with constant f and constant D that term has been zero, and the potential vorticity has been constant *at each point*. However, in the presence of a gradient of the potential vorticity preexisting in the absence of motion, the potential vorticity will not be constant at each point, even though it will be conserved following a fluid element. When the background potential vorticity is not constant, *waves may now possess nonzero potential vorticity.*

Let's try to estimate the order of magnitude of the frequency of such a wave in which the rate of change of pv is produced by motion in the field of varying ambient potential vorticity. The magnitude of the rate of change of pv can be estimated as:

- $\omega U / L$, where U is the characteristic fluid velocity in the wave and L is the characteristic horizontal scale of the wave (so that derivatives in x and y go like $1 / L$). The last term we can estimate as βU, and this yields an *estimate* of the frequency;

- $\omega = O(\beta L)$, that is, of the order of the gradient of f times the north-south excursion of the fluid element. The ratio of ω to f will then be:
- $\omega / f = O(\beta L / f_0) \ll 1$, if the beta plane approximation holds. That is, these waves, in distinction to the Poincaré waves, will have frequencies less than the Coriolis parameter; they will have time scales long compared to a day and be parametrically separated from the spectrum of gravity waves. Note, too, that this wave, again in distinction to the Poincaré and Kelvin waves, owes its very existence to the presence of rotation. We need to discover the relationship between the Rossby wave, as this β dependent wave is called, and the earlier gravity waves we have discussed. We must formulate an equation that governs both and then see how each wave type emerges from the governing equation.

To do so, it is helpful to introduce the *transport variables*

$$U = uD \tag{14.11a}$$

$$V = vD \tag{14.11b}$$

where D is the undisturbed depth. In terms of these variables,

$$U_t - fV = -gD\eta_x \tag{14.12a}$$

$$V_t + fU = -gD\eta_y \tag{14.12b}$$

$$\eta_t + U_x + V_y = 0 \tag{14.12c}$$

Cross-differentiating the momentum equations yields

$$(V_x - U_y)_t + f(U_x + V_y) + \beta V = -g(D_x\eta_y - D_y\eta_x) \tag{14.13}$$

The divergence of the momentum equations yields

$$(V_y + U_x)_t - f(V_x - U_y) + \beta U = -g\left[(D\eta_y)_y + (D\eta_x)_x\right] \tag{14.14}$$

We can eliminate the vorticity-like term between the two equations by taking the time derivative of the divergence equation and adding to it the vorticity equation multiplied by f to obtain

$$\Re(V_y + U_x) + \beta(U_t + fV) = -g\frac{\partial}{\partial t}\nabla \cdot (D\nabla \eta) + gf J(\eta, D) \tag{14.15}$$

In the above equation, we have introduced two new operators,

$$\Re(a) \equiv \left(\frac{\partial^2}{\partial t^2} + f^2\right)a \tag{14.16a}$$

$$J(a,b) = a_x b_y - a_y b_x \tag{14.16b}$$

where a and b are arbitrary functions of x, y and t.

The equation for mass conservation allows us to eliminate the divergence of the transport to obtain

$$-\Re\eta_t + \beta[U_t + fV] = -g\nabla\cdot(D\nabla\eta_t) + gfJ(\eta,D) \tag{14.17}$$

Note that were $\beta = 0$, we would have a single equation in η, and it would in fact be the equation previously derived for gravity waves in the presence of rotation with the important exception of the derivatives of D on the right-hand side. To obtain a single equation in η, we must work a little harder.

As before, we can derive equations relating U and V to the free surface elevation. From manipulating the momentum equations we obtain, as before,

$$\begin{aligned}
\Re U &= -gD\eta_{xt} - gDf\,\eta_y\\
\Re V &= -gD\eta_{yt} + gDf\,\eta_x
\end{aligned} \tag{14.18}$$

Using these relations, we can eliminate U and V from the previous equation for η to finally obtain

$$\frac{\partial}{\partial t}\underbrace{\Re[g\nabla\cdot(D\nabla\eta)-\Re\eta]}_{1}-\underbrace{gf\Re J(\eta,D)}_{2}+\underbrace{\beta\big(-gD\eta_{xtt}\big)}_{3a}-\underbrace{2\beta gDf\eta_{yt}}_{3b}+\underbrace{\beta gDf^2\eta_x}_{3c}=0 \tag{14.19}$$

This is a single equation for the free surface height. It is valid (or should be) for both Poincaré and Rossby waves, but given that the former have frequencies greater than f and the latter have frequencies less than f, some terms in the equation may be important for one wave and not for the other.

Let's estimate the various bracketed terms in the above equation for the case of the waves that have frequencies greater than f. We will estimate each term separately and then their ratios:

$$(1)=O(\omega^3 c_0^2/L^2)\eta \qquad (2)=O(gf\omega^2 h_b/L^2)\eta$$

$$(3a)=O(\beta c_0^2\omega^2/L)\eta \qquad (3b)=O(\beta c_0^2 f\omega/L)\eta \qquad (3c)=O(\beta c_0^2 f^2/L)\eta$$

Here we have estimated the operator $\Re = O(\omega^2)$ and have used L to estimate horizontal derivatives.

The ratio $(2)/(1)$ is

$$\frac{(2)}{(1)}=\frac{fg\omega^2 h_b L^2}{L^2\omega^3 c_0^2}=\frac{f}{\omega}\frac{h_b}{D} \tag{14.21}$$

For waves that have frequencies $\geq f$, it follows that *for such waves* the second term will be small with respect to the first term, since we have assumed $h_b \ll D$. Similarly,

$$\frac{(3a)}{(1)}=\frac{\beta L}{\omega}\leq\frac{\beta L}{f}\ll 1 \tag{14.22}$$

as a consequence of the beta plane approximation.

It also follows that

$$\frac{(3b)}{(1)} = \frac{\beta L f}{\omega^2} \leq \frac{\beta L}{f} \ll 1 \tag{14.23a}$$

$$\frac{(3c)}{(1)} = \frac{\beta L}{\omega^3} f^2 \leq \frac{\beta L}{f} \ll 1 \tag{14.23b}$$

Thus, for waves whose frequencies exceed f, the governing equation within an error of the order of $(\beta L / f_0, h_b / D_0)$ remains the same equation as before; namely,

$$\Re \frac{\partial}{\partial t} \left[c_0^2 \nabla^2 \eta - \frac{\partial^2 \eta}{\partial t^2} - f_0^2 \eta \right] = 0 \tag{14.24}$$

so that we will obtain the same Poincaré and Kelvin waves as before and the new terms in the governing equation will give rise, at most, to small corrections to the frequency and structure (one might be interested in finding those corrections but our main conceptual point here is that they *are* just corrections to the basic rotational-gravity waves we have already found).

On the other hand, for $\omega \ll f$, the balance of terms will be quite different. For example, the operator $\Re = O(f^2)$ and each term can be estimated to have the order

$$(1) = (\omega f^2 c_0^2 / L^2) \eta \qquad (2) = (f^3 g h_b / L^2) \eta$$

$$(3a) = (\beta c_0^2 \omega^2 / L) \eta \qquad (3b) = (\beta c_0^2 \omega f / L) \eta \qquad (3c) = (\beta c_0^2 f^2 / L) \eta$$

Therefore, for frequencies in the range of the estimated Rossby wave frequency,

$$\frac{(2)}{(1)} = \frac{f}{\omega} \frac{h_b}{D} = O(1)$$

$$\frac{(3a)}{(1)} = \frac{\beta L}{f} \frac{\omega}{f} \ll 1$$

$$\frac{(3b)}{(1)} = \frac{\beta L}{f} \ll 1 \tag{14.25}$$

$$\frac{(3c)}{(1)} = \frac{\beta L}{\omega} = O(1)$$

so that for low frequency motions, the approximation to the governing equation is

$$f^2 \frac{\partial}{\partial t} \left[c_0^2 \nabla^2 \eta - f^2 \eta \right] - f^3 g J(\eta, D) + \beta g D f^2 \eta_x = 0 \tag{14.26a}$$

or

$$\frac{\partial}{\partial t} \left[\nabla^2 \eta - \frac{f^2}{c_0^2} \eta \right] + \beta \eta_x + \frac{f}{D_0} J(\eta, h_b) = 0 \tag{14.26b}$$

Recalling that $f = f_0 + \beta y$ and that the second term is much smaller than the first, we have a uniform approximation to the above equation as

$$\frac{\partial}{\partial t}\left[\nabla^2 \eta - \frac{f_0^2}{c_0^2}\eta\right] + J(\eta, f + f_0 h_b / D_0) = 0 \tag{14.27}$$

where we have also used the smallness of h_b with respect to D_0. Note that in this equation, supposedly valid for low frequency waves, f and D are considered constants except in places where they are spatially differentiated. We will have to work a little harder to justify this heuristic derivation, but the outlines of the scaling justification should be clear at this point.

Let's look for plane wave solutions, and to make the notation simple to begin with let's examine the simple case where h_b is a function only of y and such that its derivative with respect to y is constant, i.e., a constant bottom slope. Plane waves of the form

$$\eta = A e^{i(kx+ly-\omega t)} \tag{14.28}$$

will be a solution of the above equation, if

$$\omega = -\frac{\hat{\beta}k}{k^2 + l^2 + f^2/c_0^2} \quad \text{(Rossby wave dispersion relation).} \tag{14.29}$$

Here we have defined

$$\hat{\beta} = \beta + f_0 \frac{\partial h_b}{\partial y} \tag{14.30}$$

as an "effective β."

There are a number of astonishing properties of the dispersion relation. The dispersion relation itself is shown in Fig. 14.3.

In Fig. 14.3, the frequency is scaled by $\beta c_0 / f_0$ and the wave numbers by the deformation radius c_0 / f_0, and the y wave number has been chosen to be f_0 / c_0.

First of all and most striking is that for each positive k *there is only one value of ω, and it is always < 0*. So, the phase speed of Rossby waves is always towards negative x (in this case where the bottom slope is in the y-direction and where we assume that $\hat{\beta} > 0$). The topographic slope could have the opposite sign, and the wave could have its phase propagate to positive x, but the important thing is that there is only one value for the phase speed. Previously, for all the gravity waves we have studied, for each wave propagating to the right, there was one propagating to the left with a frequency of the same magnitude. This is *not* true for the Rossby wave. Space is no longer dynamically isotropic. The dynamics recognizes, for example, which way north is by the direction of the increase of f. Once there is a special direction in space picked out for the wave, all its properties will manifest that non-isotropy.

The phase of the wave propagates in such a way that an observer, riding on the wave crests and looking in the direction of propagation, would see higher ambient potential vorticity on his right.

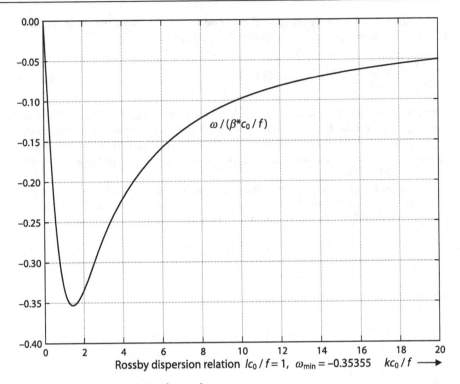

Fig. 14.3. The dispersion relation for Rossby waves

The maximum frequency (numerically) as a function of k will occur when

$$k = \left(l^2 + f_0^2 / c_0^2\right)^{1/2} \tag{14.31a}$$

$$-\omega_{max} = \frac{\hat{\beta}}{2\sqrt{l^2 + f_0^2 / c_0^2}} \tag{14.31b}$$

Over both k and l, the maximum frequency will occur when the y wave number is zero (i.e., when the y wavelength is very much less than the deformation radius) so that the overall maximum of the Rossby wave frequency and thus the minimum of the Rossby wave period is

$$-\omega_{max(all\ k,l)} = \frac{\hat{\beta} c_0}{2 f_0} \tag{14.32}$$

Second, it is dynamically impossible at this level of approximation to distinguish between the effect of the Earth's sphericity and the effect of a uniform bottom slope on a *flat* Earth in providing the necessary ambient potential vorticity gradient to support the Rossby wave. This fact has often been used with profit to construct laboratory models of oceanic waves or circulations that depend on the effect of β by instead introducing a uniformly sloping bottom.

The fact that the phase speed is always directly to the west (if we think about the planetary beta factor) seems puzzling. What would happen to wave energy in a basin if it always moved westward and never had a chance to return eastward? Of course, we are already alert to the fact that the energy moves with the group velocity and not the phase speed, so it is important to calculate the group velocity.

From the dispersion relation (and for now we will simply write β for the effective ambient pv gradient),

$$c_{gx} = \beta \frac{k^2 - (l^2 + f_0^2 / c_0^2)}{\left[k^2 + l^2 + f_0^2 / c_0^2 \right]^2} \tag{14.33a}$$

$$c_{gy} = \beta \frac{2kl}{\left[k^2 + l^2 + f_0^2 / c_0^2 \right]^2} \tag{14.33b}$$

The group velocity in the x-direction is of two signs, although the phase speed is always negative. For $k^2 > l^2 + f_0^2 / c_0^2$, i.e., for waves that are short in the x-direction, the group velocity component in the x-direction is positive, while for long waves in x, that is, $k^2 < l^2 + f_0^2 / c_0^2$, the group velocity is negative, i.e., westward. Long waves have their energy propagating westward, and short waves have their energy propagating eastward.

Figure 14.4 shows the group velocity in the x-direction as a function of k (scaled with the deformation radius).

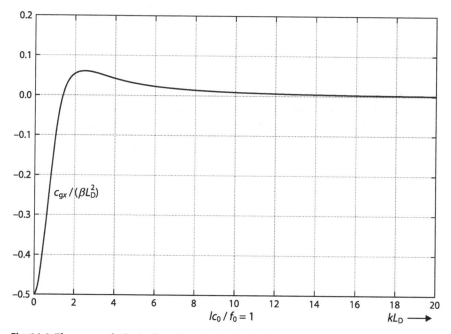

Fig. 14.4. The group velocity in the x-direction for the Rossby wave

The group velocity in the x-direction is, of course, zero when $k^2 = l^2 + f_0^2/c_0^2$, and it has its positive maximum at

$$k = \sqrt{3}\left[l^2 + f_0^2/c_0^2\right]^{1/2}$$

That maximum positive group velocity in the x-direction is

$$c_{gxmax} = \frac{\beta}{8\left[l^2 + f_0^2/c_0^2\right]} \tag{14.34}$$

while its minimum group velocity, or equivalently, its maximum *negative* group velocity occurs at $k = 0$, the longest waves in x, and is equal to

$$c_{gxmin} = -\beta\frac{c_0^2}{f_0^2} = -\beta L_D^2 \tag{14.35}$$

where we have introduced the notation $L_D \equiv c_0/f_0$ for the deformation radius. Note that the maximum speed to the west is eight times greater than the speed to the east. The westward moving long waves have a much swifter speed of energy propagation than the shorter waves, whose energy moves eastward.

The group velocity in the y-direction can have either sign, depending on the sign of the product of k and l. Note that since

$$c_y = \frac{\omega}{l} \tag{14.36a}$$

$$c_{gy}/c_y = -\frac{2l^2}{(k^2 + l^2 + 1/L_D^2)} < 0 \tag{14.36b}$$

the group velocity in the y-direction is oppositely directed to the phase speed in the y-direction. This is reminiscent of the oppositely opposed phase and group speeds in the vertical direction for internal gravity waves, and it is left to the student to develop and complete the analogy.

Rossby Waves (*Continued*), Quasi-Geostrophy

For the Poincaré wave, $\omega \geq f$, and so the wave motion is not in geostrophic balance, while for the Rossby wave,

$$\omega \leq \beta L_d = \beta c_0 / f \tag{15.1}$$

so that

$$\omega / f = \frac{\beta L_d}{f} \leq \frac{\beta L}{f} \tag{15.2}$$

where L is the scale of the motion. Thus, for Rossby waves, the frequency is less than f so that in the x-momentum equation, for example,

$$\overset{1}{\underset{\smile}{u_t}} - \overset{2}{\underset{\smile}{fv}} = -g\eta_x \tag{15.3}$$

term (1) will be less than term (2) by the order of ω / f. The velocity will be in approximate geostrophic balance to that order. This is similar to the hydrostatic approximation in which the vertical pressure gradient can be calculated *as if* the fluid were at rest, even though it is motion, because the vertical accelerations are very small when the aspect ratio D/L of the motion is small. Here the *horizontal* pressure gradient is given by the Coriolis acceleration *as if* there were no acceleration of the relative velocity, i.e., as if the flow were uniform in space and time even though it is not because that acceleration is very small compared to the Coriolis acceleration.

These simple intuitive ideas form the basis of a formal theory, *quasi-geostrophy*, that systematizes that idea (Pedlosky 1987). The reason why we have to be formal is that otherwise (and brutal historical experience shows the foolishness of taking the careless path) it is not clear how to proceed in the approximation beyond its initial step to arrive at an equation of motion that is dynamically consistent and conserves in appropriate approximate form *all* the conservation principles present in the original, more complex set of equations. We want the simplified set so that we can penetrate more deeply into the low frequency limit of the dynamics, which is of special interest in oceanography and meteorology, but we clearly want to do it right. It is always easy to do it wrong.

For example, if the Coriolis parameter varies, where can that variation be ignored and where must it be maintained? The same question will hold with regard to the variation of the depth, which we saw in the last lecture acted dynamically similarly to the β-effect. If the motion is in geostrophic balance at the lowest order, how can we consistently calculate its evolution in time or its structure in space? Geostrophy only tells us that if we know the velocity, we can calculate the pressure, or vice-versa, but it does not tell us how to calculate either of them from initial or boundary data. Our task now is to take up this question, and our goal is to derive a set of equations for the low frequency motion of the fluid, in this example a homogeneous layer of fluid, that is simpler than the initial set but rich enough to allow us to go beyond the investigation of simple plane wave theory.

To do so, we must bring to the analysis certain physical ideas. Nothing here is, to begin with, automatic. Based on our experience, we describe a set of consistent presumptions and find the dynamics consistent with those presumptions. If those a priori ideas are valid and physically interesting, the resulting equations will give us interesting results; otherwise, they will not.

We presume, *a priori* that the time scales of the motions of interest are long compared to $1/f$. Or, more formally, *if T* (think of a wave period) is the time scale of the motion such that

$$\frac{\partial}{\partial t} = O(1/T) \tag{15.4}$$

then we presume

$$fT \gg 1, \quad \text{i.e.} \quad (\omega/f) \ll 1 \tag{15.5}$$

We also assume that there is a length scale, L, which characterizes the horizontal scale of the motion such that horizontal derivatives can be estimated by $1/L$. Further, we assume that there is a scale for the fluid velocity, U, which characterizes the motion of the *fluid*. This means that the nonlinear part of the total derivative, i.e., terms like uv_x, will be of the order of U^2/L, and this introduces an *advective time scale*, $T_{\text{advective}} = L/U$. The condition that the advective time scale be long compared to the rotation period is $fT_{\text{advective}} \gg 1$ or equivalently that

$$\varepsilon \equiv \frac{U}{fL} \ll 1 \tag{15.6}$$

where ε is the *Rossby number*. Actually, we will define the Rossby number in what follows in terms of the constant value of f at the reference latitude so that

$$\varepsilon = \frac{U}{f_0 L} \tag{15.7}$$

will be a constant.

We need to carefully estimate *all* the terms in the equations of motion and obtain an easy way to keep track of their relative sizes. That is most efficiently done by introducing non-dimensional variables. These non-dimensional variables will be $O(1)$ if we have chosen the scale for time, length and velocity correctly for the motion of interest.

We introduce non-dimensional variables as follows; they will *temporarily be denoted by primes.*

$$(x, y) = L(x', y')$$
(15.8a)

$$(u, v) = U(u', v')$$
(15.8b)

$$t = Tt'$$
(15.8c)

We must also scale the free surface height η. How should we do that in a way that is consistent with our scales for velocity, length, and time and our presumption that the motion is of low frequency? We *anticipate* that the motion to the lowest order will be both hydrostatic and geostrophic (almost; that is where the *quasi* comes in) so that to the lowest order we expect that

$$g\nabla\eta = O(f_0 U)$$
(15.9)

but if our estimates of spatial scale are correct,

$$g\nabla\eta = O(g\eta / L)$$
(15.10)

therefore,

$$\eta = O\left(\frac{f_0 UL}{g}\right)$$
(15.11)

and so

$$\frac{\eta}{D_0} = O\left(\frac{f_0 UL}{gD_0}\right) = \varepsilon \frac{f_0^2 L^2}{gD_0} = \varepsilon\left(\frac{L}{L_D}\right)^2$$
(15.12)

where the deformation radius is, as it was defined earlier,

$$L_D = c_0 / f_0 = (gD_0)^{1/2} / f_0$$
(15.13)

For motions whose lateral scale is of the order of the deformation radius, we can expect that the proportional change in layer thickness due to the motion, i.e., η / D will be of the order of the Rossby number and hence small. If L is much larger than the deformation radius, we may still be able to consider the proportional change in layer thickness due to the motion as small, if the product is

$$\left(\frac{f_0 UL}{gD_0}\right) = \varepsilon\left(\frac{L}{L_D}\right)^2 \ll 1$$

i.e., for a small enough ε.

We therefore formally introduce the scaling for the free surface height,

$$\eta = D_0 \varepsilon F \eta' \tag{15.14a}$$

$$F \equiv \frac{f_0^2 L^2}{g D_0} = \left(\frac{L}{L_D}\right)^2 = O(1) \tag{15.14b}$$

This allows us to write for the total thickness of the fluid

$$D = D_0 \left(1 + \varepsilon F \eta' - \frac{h_b}{D_0}\right) \tag{15.15}$$

and we will assume throughout that

$$\frac{h_b}{D_0} \ll 1$$

Similarly, with the beta-plane approximation in the representation of f, we take

$$f = f_0(1 + \beta y / f_0) = f_0 \left((1 + \frac{\beta L}{f_0} y')\right) \tag{15.16}$$

and we assume that the parameter $\beta L / f_0 \ll 1$.

Now we insert each of these relations into the equations of motion; for example, in the x-momentum equation we have

$$\frac{U}{T} \frac{\partial u'}{\partial t'} + \frac{U^2}{L} \left[u' u'_{x'} + v' u'_{y'}\right] - f_0 \left(1 + \frac{\beta L}{f_0} y'\right) v' = -f_0 U \eta'_{x'} \tag{15.17}$$

Dividing by the factor $f_0 U$ yields

$$\varepsilon_T \frac{\partial u'}{\partial t'} + \varepsilon \left[u' u'_{x'} + v' u'_{y'}\right] - \left(1 + \frac{\beta L}{f_0} y'\right) v' = -\eta'_{x'} \tag{15.18}$$

where

$$\varepsilon_T = \frac{1}{f_0 T} \ll 1 \tag{15.19}$$

Similarly for the y-momentum equation we obtain

$$\varepsilon_T \frac{\partial v'}{\partial t'} + \varepsilon \left[u' v'_{x'} + v' v'_{y'}\right] + \left(1 + \frac{\beta L}{f_0} y'\right) u' = -\eta'_{y'} \tag{15.20}$$

The conservation of mass equation in dimensional units is

$$\eta_t + (uD)_x + (vD)_y = 0$$

and by inserting the scaling variables above and the form for D, i.e.,

$$D = D_0\left(1 + \varepsilon F\eta' - \frac{h_b}{D_0}\right)$$

we obtain

$$\frac{D_0}{T}\varepsilon F\eta'_{t'} + D_0\frac{U}{L}\left[\{u'(1+\varepsilon F\eta'-h_b/D_0)\}_{x'} + \{v'(1+\varepsilon F\eta'-h_b/D_0)\}_{y'}\right] = 0 \qquad (15.21a)$$

or

$$\varepsilon_T F\eta'_{t'} + \vec{u}'\cdot\nabla'(\varepsilon F\eta'-h_b/D_0) + (1+\varepsilon F\eta'-h_b/D_0)\nabla\cdot\vec{u}' = 0 \qquad (15.21b)$$

At this point, our equations look as if they have come down with a bad case of acne; the primes make the equations look very ugly. The traditional thing to do at this point is to improve the aesthetic quality of the development by *dropping primes*. Henceforth, *unless otherwise noted*, unprimed variables will be *non-dimensional*, and we will use asterisks to denote dimensional variables, e.g., $x^* = Lx$.

Our dynamical system of equations can now be neatly written as

$$\varepsilon_T\frac{\partial\vec{u}}{\partial t} + \varepsilon\vec{u}\cdot\nabla\vec{u} + \left(1+\frac{\beta L}{f_0}y\right)\hat{z}\times\vec{u} = -\nabla\eta \qquad (15.22a)$$

$$\varepsilon_T\frac{\partial\eta}{\partial t} + \vec{u}\cdot\nabla(\varepsilon F\eta-h_b/D_0) + (1+\varepsilon F\eta-h_b/D_0)\nabla\cdot\vec{u} = 0 \qquad (15.22b)$$

These equations contain several small parameters. There are the two Rossby numbers ε_T and ε, as well as a measure of the sphericity factor $\beta L/f_0$, and of course, the deviation of the rest thickness of the layer in absence of motion from the constant measured by h_b/D_0. We will assume that the parameter F is order one, i.e., that the horizontal length scale is of the order of the deformation radius. We will expand the equations of motion in an asymptotic series in ε and assume that each of the Rossby numbers is of the same order, that is, that

$$\frac{\varepsilon}{\varepsilon_T} = \frac{UT}{L} = U/c = O(1) \qquad (15.23)$$

and if we want to subsequently linearize the resulting equations, we can assume at the end of our labors that this parameter is small with respect to one. We will also assume that $\beta L/f_0$ is of order ε, which implies that

$$\frac{\varepsilon}{\beta L/f_0} = \frac{U}{\beta L^2} = O(1) \qquad (15.24)$$

The Rossby number, ε, is a ratio of the relative vorticity, of the order U/L to the planetary vorticity, f. It is assumed small. The ratio above $U/\beta L^2$ is the ratio of the relative vorticity *gradient* to the planetary vorticity *gradient* and that can be order one. That is because the relative vorticity varies relatively fast on the scale L, while the planetary vorticity varies more slowly on the scale of the Earth's radius, R. The fact is that $L / R \ll 1$ is a requirement of the beta-plane approximation.

So we expand each variable in the series:

$$\vec{u}(x,y,t,\varepsilon) = \vec{u}_0(x,y,t) + \varepsilon \vec{u}_1(x,y,t) + \ldots \tag{15.25a}$$

$$\eta(x,y,t,\varepsilon) = \eta_0(x,y,t) + \varepsilon \eta_1(x,y,t) + \ldots \tag{15.25b}$$

Note that each subscripted variable is *independent* of ε. Thus, when this series is inserted in the equations of motion, like orders in ε must balance for the equations to be valid for ε small, but arbitrary. This leads to the following set of equations.

Collecting the $O(1)$ terms in the momentum equation,

$$\hat{z} \times \vec{u}_0 = -\nabla \eta_0 \tag{15.26}$$

or in component form,

$$u_0 = -\eta_{0y} \tag{15.27a}$$

$$v_0 = \eta_{0x} \tag{15.27b}$$

which is simply the geostrophic balance at the lowest order (note that the variation of the Coriolis parameter does *not* enter at this order; it *as if f* were constant in the lowest order geostrophic balance). Note that as a consequence of geostrophy,

$$\frac{\partial u_0}{\partial x} + \frac{\partial v_0}{\partial y} \equiv 0 \tag{15.28}$$

The geostrophic velocity, with constant f, is horizontally nondivergent.

When we look for order one terms in the mass conservation equation, the development depends on whether h_b / D_0 is order one or less. We will assume it is $O(\varepsilon)$ so that, as the beta effect, that variation does not enter at the lowest order. It is left for the student to discover what the dynamics will look like if h_b / D_0 is $O(1)$. If h_b / D_0 is $O(\varepsilon)$, then all terms in the equation for η are of order ε, noting that the horizontal divergence of the $O(1)$ velocity vanishes.

We then are left at order one, with only the diagnostic relation between the pressure gradient and the geostrophic velocity with no way, at this order, to calculate the evolution of the fields, e.g., to discuss Rossby waves. We must go to a higher order in our expansion to do so. It is precisely for this reason that higher order small terms must be considered and that we must be exquisitely careful to consider all small terms that are of the same order. It is for this reason that we have gone through the scaling and the non-dimensionalization so that we can be sure we are not leaving a small term out while considering others. We need to keep the dynamics consistent if the final result is to be physically sensible.

At $O(\varepsilon)$, i.e., keeping terms of $O(\varepsilon)$, the conservation of mass and momentum equations yield

$$\frac{\varepsilon_T}{\varepsilon}F\frac{\partial \eta_0}{\partial t}+F\vec{u}_0\cdot\nabla\eta_0-\vec{u}_0\cdot\nabla\left(\frac{h_b}{\varepsilon D_0}\right)+\nabla\cdot\vec{u}_1=0 \tag{15.29a}$$

$$\frac{\varepsilon_T}{\varepsilon}\frac{\partial u_0}{\partial t}+u_0u_{0x}+v_0u_{0y}-v_1-y\frac{\beta L^2}{U}v_0=-\eta_{1x} \tag{15.29b}$$

$$\frac{\varepsilon_T}{\varepsilon}\frac{\partial v_0}{\partial t}+u_0v_{0x}+v_0v_{0y}+u_1+y\frac{\beta L^2}{U}u_0=-\eta_{1y} \tag{15.29c}$$

Here we have used the fact that

$$\frac{\beta L}{\varepsilon f_0}=\frac{\beta L^2}{U}=O(1)$$

and that

$$\frac{h_b}{\varepsilon D_0}=O(1)$$

in identifying terms of order ε. Note that these $O(\varepsilon)$ equations describe the rate of change with time of the $O(1)$ velocities and free surface elevation. However, the equations contain the $O(\varepsilon)$ variables as well, and so the system does not seem closed at this order. This is a little bit worrisome. Let us press on, though, by eliminating the $O(\varepsilon)$ free surface elevation from the momentum equations by cross differentiating. Using the fact that the $O(1)$ velocities have zero divergence, we obtain an equation for the evolution of the relative vorticity:

$$\frac{\varepsilon_T}{\varepsilon}\zeta_{0t}+u_0\zeta_{0x}+v_0\zeta_{0y}+\frac{\beta L^2}{U}v_0=-(v_{1y}+u_{1x}) \tag{15.30}$$

where $\zeta_0=v_{0x}-u_{0y}$ is the relative vorticity.

The interpretation of the above equation is rather interesting. The left-hand side of the equation yields the total rate of change of the sum of the relative plus planetary vorticity following a fluid element; in *dimensional variables* this would just be

$$\frac{d\zeta+f}{dt}$$

The right-hand side of the equation is minus the product of the Coriolis parameter at the reference latitude, f_0 and the divergence, i.e., $-f_0(u_x+v_y)$. In our scaling universe, the Coriolis parameter is order one, and the horizontal divergence is $O(\varepsilon)$, so the product is of the same order as the rate of change $O(\varepsilon)$ of the order one relative vorticity. Note that this source of vorticity normally would contain the convergence not only in the presence of the reference Coriolis parameter but of the full vorticity $f + \zeta$. However, those corrections are of a higher order in Rossby number, it would not be consistent to keep them, and indeed, they do not appear in the non-dimensional vorticity equation we have derived. This is one of the advantages of the careful bookkeeping that the method does for us.

We still are in some difficulty, apparently, because the rate of change of the relative vorticity is given by the higher order divergence, which we don't know. We can eliminate the divergence, though, through the use of the equation for mass conservation. Thus,

$$\frac{d_0}{dt}(\zeta_0 - F\eta_0) + \vec{u}_0 \cdot \nabla\left\{\hat{\beta}y + h_b/\varepsilon D_0\right\} = 0 \qquad (15.32)$$

We have defined

$$\frac{d_0}{dt} \equiv \frac{\varepsilon_T}{\varepsilon}\frac{\partial}{\partial t} + (\vec{u}_0 \cdot \nabla) \qquad (15.33a)$$

$$\hat{\beta} = \frac{\beta L^2}{U} \qquad (15.33b)$$

Indeed, the equation can be written in conservation form more simply as

$$\frac{d_0}{dt}\left[\zeta_0 - F\eta_0 + \hat{\beta}y + h_b/\varepsilon D_0\right] = 0 \qquad (15.34)$$

All variables in the above equation are $O(1)$, and the equation is a conservation equation for an $O(1)$ variable. What is that quantity? By now you should have the feeling from its form that is the potential vorticity or some suitable approximation to it valid for a small Rossby number. We shall check that shortly, but first we need to make a very important point. The equation is a *single* equation in several variables, ζ_0, η_0 and the two velocity components. However, the $O(1)$ geostrophic relation allows us to write all the variables in terms of the free surface elevation, since

$$u_0 = -\eta_{0y} \qquad (15.35a)$$

$$v_0 = \eta_{0x} \qquad (15.35b)$$

$$\zeta_0 = v_{0x} - u_{0y} = \eta_{0xx} + \eta_{0yy} = \nabla^2\eta_0 \qquad (15.35c)$$

Noting that the lowest order free surface elevation plays the role of a stream function for the $O(1)$ geostrophic velocities, we define

$$\psi = \eta_0 \qquad (15.36)$$

in terms of which the above equation can be rewritten:

$$\frac{d_0}{dt}\left[\nabla^2\psi - F\psi + \hat{\beta}y + h_b/\varepsilon D_0\right] = 0 \tag{15.37a}$$

or

$$\frac{\varepsilon_T}{\varepsilon}\frac{\partial}{\partial t}\left[\nabla^2\psi - F\psi\right] + \psi_x\frac{\partial}{\partial y}\left[\nabla^2\psi - F\psi\right] - \psi_y\left[\nabla^2\psi - F\psi\right]$$
$$+ \psi_x\left[\hat{\beta}y + h_b/\varepsilon D_0\right]_y - \psi_y\left[\hat{\beta}y + h_b/\varepsilon D_0\right]_x = 0 \tag{15.37b}$$

Thus, we have attained a governing equation in the single variable ψ. We introduce the notation for the *Jacobian*, $J(a,b)$, of any two functions a and b:

$$J(a,b) \equiv a_x b_y - a_y b_x$$

The equation of motion is thus

$$\frac{\varepsilon_T}{\varepsilon}\frac{\partial}{\partial t}\left[\nabla^2\psi - F\psi\right] + J(\psi, \nabla^2\psi - F\psi + \hat{\beta}y + h_b/\varepsilon D_0) = 0 \tag{15.38}$$

This equation forms the heart of our analysis of quasi-geostrophic motion, but before we proceed to its analysis and in particular its role in wave theory, it is useful to understand the origin of the equation in a more heuristic manner than our careful asymptotic derivation.

For a single layer of fluid, in *dimensional units*, the equation for conservation of potential vorticity, assuming only that the motion is hydrostatic and the horizontal velocities are independent of z, is

$$\frac{d}{dt}\left[\frac{\zeta + f}{D}\right] = 0 \tag{15.39a}$$

$$D = D_0 + \eta - h_b \tag{15.39b}$$

D is the total depth, and it departs from a constant value by a small amount; indeed, we can approximate the potential vorticity

$$q = \frac{\zeta + f}{D} = \frac{\zeta + f_0 + \beta y}{D_0 + \eta - h_b} \approx \frac{1}{D_0}(\zeta + f_0 + \beta y)(1 - \{\eta - h_b\}/D_0 + \ldots) \tag{15.40}$$

using the expansion for $1/(1+\varepsilon) = 1 - \varepsilon + O(\varepsilon)^2$.

Keeping only those terms in the above product that are of either order one or of the same order as the relative vorticity, we get

$$q \approx \frac{f_0}{D_0} + \frac{\left[\zeta + \beta y - f_0\eta/D_0 + f_0 h_b/D_0\right]}{D_0} \tag{15.41}$$

The first term in this approximation for q is an irrelevant constant. The conservation equation then applies to the second term, which aside from a multiplicative constant yields

$$\frac{d}{dt}\left(\zeta + \beta y - \frac{f_0 \eta}{D_0} + \frac{f_0 h_b}{D_0}\right) = 0 \tag{15.42}$$

At the same level of approximation, the geostrophic relation yields, *in dimensional units,*

$$u = -\frac{g}{f_0}\eta_y \tag{15.43a}$$

$$v = \frac{g}{f_0}\eta_x \tag{15.43b}$$

Thus, the pv equation becomes

$$\frac{g}{f_0}\frac{d}{dt}\left[\nabla^2 \eta - \frac{f_0^2 \eta}{gD_0} + \frac{\beta f_0 y}{g} + \frac{f_0^2 h_b}{gD_0}\right] = 0 \tag{15.44}$$

Defining (remember these are in *dimensional units*)

$$\psi = \frac{g}{f_0}\eta \tag{15.45}$$

the potential vorticity equation becomes

$$\frac{dq}{dt} \approx \frac{d}{dt}\left[\nabla^2 \psi - \frac{1}{L_D^2}\psi + \frac{f_0 h_b}{D_0} + \beta y\right] = 0 \tag{15.46}$$

If we were to scale x and y as we did earlier in the lecture, the above equation would become the dimensionless dynamical equation we obtained earlier in our more careful scaling and asymptotic expansion method. That care allowed us to be sure that in our estimate of potential vorticity we included all the correct terms (and no more) and that we could replace f with its reference value in the geostrophic relation and definition of stream function.

The end result is that *for low frequency motions with a small Rossby number, the governing equation of motion is the potential vorticity equation in which all terms are evaluated using their hydrostatic and geostrophic approximations in terms of the pressure field, in this case, the free surface height.*

Note that the total derivative *in dimensional units* is

$$\frac{d}{dt} = \frac{\partial}{\partial t} + \frac{\partial \psi}{\partial x}\frac{\partial}{\partial y} - \frac{\partial \psi}{\partial y}\frac{\partial}{\partial x} \tag{15.47}$$

As we remarked earlier, for quasi-geostrophic motions the role of the bottom slope mimics that of the beta effect, and one can use a constant bottom slope in the laboratory to model the dynamical effect of the Earth's sphericity. We can see, perhaps more easily from the non-dimensional form of the equation, that the relative importance of the beta effect and the bottom slope depend on the ratios

$$\frac{\beta L^2}{U}, \quad \frac{f_0 h_b}{U/L}$$

The first is the ratio of the planetary vorticity *gradient* to the relative vorticity *gradient*, and that ratio is typically unity, although as the horizontal scale grows larger the beta effect tends to dominate, because for a given scale for U, the relative vorticity and its gradient decrease with L. The topographic term can also be thought of as a ratio of the contribution by the topography to the potential vorticity gradient $(f_0 h_b/D_0)/L$ to the relative vorticity gradient U/L^2, and again as L increases, the topographic term tends to dominate (Note that the *equivalent* topographic beta is of the order

$$\beta_{\text{topog}} = \frac{f_0 h_b}{D_0 L}$$

so the ratio given above can be written

$$\frac{\beta_{\text{topog}} L^2}{U}$$

to complete the analogy). These terms are the contributions to the *ambient potential vorticity*, that is, the potential vorticity preexisting in the absence of any motion. When the ambient potential vorticity dominates, i.e., when the above ratios become very large, the first approximation to the potential vorticity equation is just

$$J\left(\psi, \beta y + \frac{f_0 h_b}{D_0}\right) = 0 \tag{15.48}$$

The stream function must then be constant along lines of constant ambient pv in the x-y-plane, which is an extraordinarily strong constraint. Breaking that constraint requires either a source of potential vorticity that will nudge fluid elements off the lines of constant ambient pv or regions in which dissipation (hitherto ignored) or nonlinearity become important. These considerations are of vital importance in the theory of the general circulation of the ocean, but pursuing them further here would divert us from our goal of understanding the physics of waves. Therefore, we return to the *quasi-geostrophic potential vorticity equation (qgpve)* given above. We will work in dimensional units, although our systematic derivation leaned heavily on our scaling and asymptotic approximations for a small Rossby number.

Quasi-Geostrophic Rossby Waves

We return to the qgpve and examine the nature of Rossby waves in the presence of an ambient potential vorticity gradient. For *simplicity*, we will take the ambient gradient to be a constant. We define the ambient pv as

$$Q = \beta y + f_0 h_b/D_0 \tag{15.49}$$

and assume its gradient is constant. The *linearized* form of the qgpve is then

$$\frac{\partial}{\partial t}\left[\nabla^2\psi - a^2\psi\right] + \psi_x Q_y - \psi_y Q_x = 0 \tag{15.50}$$

The quantity a^2 is $1/L_D^2$ and has the dimensions of a wave number (squared). With the above assumptions on Q, the equation has constant coefficients, and if we consider an infinite region, we can find plane wave solutions in the form

$$\psi = Ae^{i(kx+ly-\omega t)} \tag{15.60}$$

which requires that

$$-i\omega(-k^2 - l^2 - a^2) + ikQ_y - ilQ_x = 0$$

$$\Rightarrow \quad \omega = -\frac{\left[Q_y k - Q_x l\right]}{K^2 + a^2} = -\hat{z}\cdot\left[\frac{\vec{K}\times\nabla Q}{K^2 + a^2}\right] \tag{15.61}$$

$$= -\vec{K}\cdot\left[\frac{\hat{z}\times\nabla Q}{K^2 + a^2}\right]$$

where $\vec{K}$ is the two-dimensional wave vector, $K^2 = k^2 + l^2$, while $\hat{z}$ is the vertical unit vector. Finally, the frequency can be written

$$\omega = \frac{\vec{K}\cdot\left(\hat{z}\times\nabla Q\right)}{K^2 + a^2} \tag{15.62}$$

So the frequency depends on the projection of the wave vector on the direction perpendicular to ∇Q, i.e., it depends on the projection of the wave vector along the lines of constant ambient pv. Since the geostrophic velocity is perpendicular to the wave vector (why is this so?), the frequency depends on the degree to which fluid elements cross ambient pv contours. If the fluid flows along pv contours, i.e., if $\vec{u} \cdot \nabla Q = 0$, the time derivative in the linearized qgpve would be zero: no wave (Fig. 15.1).

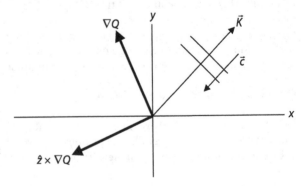

Fig. 15.1.
The relation between the wave
vector, the ambient pv gradient
and the direction of the phase
speed

Note that the phase speed in the direction of the wave vector is

$$\vec{c} = \frac{\omega}{K} \frac{\vec{K}}{K}$$

and the inner product of this pseudo vector with the vector $\hat{z} \times \nabla Q$ yields

$$\vec{c} \cdot (\hat{z} \times \nabla Q) = \frac{\left[\vec{K} \cdot \{\hat{z} \times \nabla Q\}\right]^2}{K^2 + a^2} \geq 0 \tag{15.63}$$

Thus, the phase *always* moves in the direction such that an observer riding on a crest will see larger values of Q on her right. As discussed earlier, the phase propagation is in a single direction, essentially such as to make an acute angle with the isopleths of ambient pv and to be guided by its gradient.

The case of a flat bottom arose when $\nabla Q = \beta \hat{j}$ was discussed in the last lecture. What we need to do now is to develop a clear picture of the direction and magnitude of energy propagation in the Rossby wave. This is rendered a bit tricky because of the disconcerting fact that the obvious candidate for the energy flux at the lowest order in Rossby number $p\vec{u}$ is horizontally nondivergent. That is, if the velocity is geostrophic, both its divergence and its inner product with grad p vanish identically. That is not a useful tool for calculating the transfer of energy. The difficulty is only resolved by noting that as in the case of the dynamics, the energy flux will involve the $O(\varepsilon)$ contributions of the pressure to calculate its gradient and $O(\varepsilon)$ contributions to the velocity to calculate the velocity's horizontal divergence. That awkwardness can be avoided by dealing directly with the qgpve, as we shall show in the next lecture.

Energy and Energy Flux in Rossby Waves

In discussing the energy and its flux for Rossby waves, we encounter the problem that the natural definition of the energy flux at the lowest order $p\bar{u}$ is horizontally non-divergent and therefore has no effect on the wave energy. To discuss the real energy flux, one has to include the divergent, non-geostrophic $O(\varepsilon)$ part of the velocity field as well as the pressure contribution at this order. This would be a messy business, and what is worse is that the solution of the quasi-geostrophic potential vorticity equation doesn't give us these quantities as part of the solution. Is there a way we can describe the energy flux entirely within the quasi-geostrophic framework? The answer is yes, and it follows from a direct consideration of the linear quasi-geostrophic equation. First, though, let us orient the y-axis in the direction of the gradient of the ambient potential vorticity, ∇Q, and call the magnitude of the gradient β for obvious reasons. As long as the gradient is a constant, there is no loss of generality. It will be up to the student to try to generalize these results when the gradient is not constant. The linear qgpve is

$$\frac{\partial}{\partial t}\left[\nabla^2\psi - a^2\psi\right] + \beta\frac{\partial\psi}{\partial x} = 0 \tag{16.1}$$

If we multiply the equation by the stream function, we obtain

$$\nabla\cdot\left(\psi\nabla\psi_t\right) - \nabla\psi\cdot\nabla\psi_t - a^2\psi\psi_t + \beta\psi\psi_x = 0 \tag{16.2}$$

which can be rewritten:

$$\frac{\partial}{\partial t}\left[\frac{(\nabla\psi)^2}{2} + \frac{a^2\psi^2}{2}\right] + \nabla\cdot\left\{-\psi\nabla\psi_t - \frac{1}{2}\hat{x}\beta\psi^2\right\} = 0 \tag{16.3}$$

where $\hat{x}$ is a unit vector in the x-direction. From the definition of the stream function, it is clear that the first term in the square brackets is the kinetic energy

$$\frac{1}{2}(u^2 + v^2)$$

which is associated with the $O(1)$ geostrophic motion. The second term is potential energy, since

$$\frac{a^2\psi^2}{2} = \frac{f_0^2}{2c_0^2}\frac{g^2\eta^2}{f_0^2} = \frac{g\eta^2}{2D_0} \tag{16.4}$$

(multiplication of the whole equation by D_0 is necessary to give the total energy in the water column, but this obvious step is trivial).

Therefore, the first term in the above equation will be the sum of the kinetic and potential energies in the wave field. The term in the curly bracket is a vector, $\vec{S}$, whose divergence alters the local wave energy. Note that $\vec{S}$ is given entirely in terms of the geostrophic stream function, ψ.

Thus, we have the usual energy flux equation:

$$\frac{\partial E}{\partial t} + \nabla \cdot \vec{S} = 0 \tag{16.5a}$$

$$\vec{S} = -\psi \nabla \psi_t - \hat{x}\beta \psi^2 / 2 \tag{16.5b}$$

To get a better feeling for the flux vector $\vec{S}$, consider a Rossby wave packet

$$\psi = A\cos(kx + ly - \omega t) \tag{16.6}$$

where the amplitude A is a slowly varying function of space and time. Let's calculate the energy. The kinetic energy is

$$KE = \frac{1}{2}\left[\psi_x^2 + \psi_y^2\right] = \frac{A^2}{2}(k^2 + l^2)\sin^2(kx + ly - \omega t) \tag{16.7}$$

and averaged over a period,

$$\langle KE \rangle = \frac{A^2}{4}(k^2 + l^2) \tag{16.8}$$

while the potential energy, which similarly averaged, is

$$\langle PE \rangle = \frac{A^2}{4}a^2 \tag{16.9}$$

Thus, the total energy in the Rossby wave averaged over a period (or wavelength) is

$$\langle E \rangle = \frac{A^2}{4}(K^2 + a^2), \quad K^2 = k^2 + l^2 \tag{16.10}$$

Now we need to calculate the flux vector $\vec{S}$. Using the notation

$$\vartheta = kx + ly - \omega t$$

$$\vec{S} = -\psi \nabla \psi_t - \beta \frac{\psi^2}{2}\hat{x}$$

$$= -A\cos(\vartheta)\{\omega \vec{K}A\cos(\vartheta)\} - \beta\frac{A^2}{2}\cos^2(\vartheta) \tag{16.11}$$

this becomes, when averaged over a wave period,

$$\langle \vec{S} \rangle = \beta \frac{A^2}{2} \frac{\vec{K}k}{(K^2+a^2)} - \beta \frac{A^2}{4} \hat{x} \tag{16.12}$$

To arrive at this, we have used the dispersion relation

$$\omega = -\frac{\beta k}{(K^2+a^2)} \tag{16.13}$$

Part of the energy flux vector is in the direction of the wave number, and a part lies along the x-axis. It is useful to decompose the flux vector into its x and y components. If $\hat{y}$ is the unit vector along the y-axis,

$$
\begin{aligned}
\langle \vec{S} \rangle &= \beta \frac{A^2}{4} \left[\hat{x} \left(\frac{2k^2 - k^2 - l^2 - a^2}{K^2+a^2} \right) + \hat{y} \left(\frac{2kl}{K^2+a^2} \right) \right] \\
&= \beta \frac{A^2}{4} (K^2+a^2) \left[\hat{x} \left(\frac{k^2 - l^2 - a^2}{(K^2+a^2)^2} \right) + \hat{y} \left(\frac{2kl}{(K^2+a^2)^2} \right) \right] \\
&= \vec{c}_g \langle E \rangle
\end{aligned}
\tag{16.14}
$$

where we have used the formula derived earlier for the group velocity of Rossby waves; namely,

$$\vec{c}_g = \beta \left[\hat{x} \frac{k^2 - l^2 - a^2}{(K^2+a^2)^2} + \hat{y} \frac{2kl}{(K^2+a^2)^2} \right] \tag{16.15}$$

This allows us to write the energy equation:

$$\frac{\partial \langle E \rangle}{\partial t} + \nabla \cdot \vec{c}_g \langle E \rangle = 0 \tag{16.16}$$

The Energy Propagation Diagram

As noted earlier, to obtain the full energy flux written in terms of the pressure work term, one would have to include the effects of the order Rossby number (ε) velocity. That velocity is *not* horizontally nondivergent. Therefore, the total velocity required for the calculation of the energy flux does not satisfy the condition that it would be perpendicular to the wave vector. That in turn implies that the group velocity will not be perpendicular to the wave vector (nor parallel to it). To discuss the relation between the wave vector's direction and the direction of the group velocity (which is, from above, the direction of the energy flux), we will employ a graphical development due originally to Longuet-Higgins (1964). Consider waves of frequency

$$\omega = -\frac{\beta k}{k^2 + l^2 + a^2} \tag{16.17}$$

We will use the convention that $k > 0$, so that ω is < 0. For a given ω, the possible locus of wave numbers in the k-l-plane satisfies

$$k^2 + l^2 + \beta \frac{k}{\omega} + a^2 = 0$$

$\Rightarrow$ (16.18)

$$\left(k - \left\{\frac{\beta}{-2\omega}\right\}\right)^2 + l^2 = \frac{\beta^2}{4\omega^2} - a^2$$

The wave vector must therefore lie on a circle (see Fig. 16.1) in the k-l-plane centered at the point

$$\left(\frac{\beta}{-2\omega}, 0\right)$$

with radius

$$\sqrt{\frac{\beta^2}{4\omega^2} - a^2}$$

Note that for the circle's radius to exist, the frequency has to less than the maximum Rossby wave frequency $\beta / 2a$.

The point of the circle closest to the origin lies on the k-axis at a distance

$$k_m = \frac{\beta}{-2\omega} - \sqrt{\frac{\beta^2}{4\omega^2} - a^2}$$ (16.19)

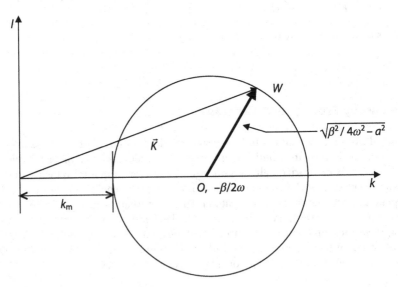

Fig. 16.1. The energy propagation diagram

When the deformation radius $c_0/f_0 = 1/a$ is very large so that $a \longrightarrow 0$, the point $k_m \longrightarrow 0$.

Now let's calculate the energy flux vector $\rangle \vec{S} \langle$. From our earlier results,

$$
\begin{aligned}
\langle \vec{S} \rangle &= \frac{A^2}{2} \vec{K} \frac{\beta k}{K^2 + a^2} - \frac{A^2 \beta}{4} \hat{x} \\
&= \frac{A^2(-\omega)}{2} \left[\vec{K} - \frac{\beta}{2(-\omega)} \right] \\
&= \frac{A^2}{2(-\omega)} \left[k - \frac{\beta}{(-2\omega)}, l \right] \\
&= \frac{A^2}{2(-\omega)} \overrightarrow{OW}
\end{aligned}
\tag{16.20}
$$

where the vector $\overrightarrow{OW}$ shown in Fig. 16.1 is directed from the origin of the wave number circle to the point on the circle corresponding to wave number $\vec{K}$. Note that the length of the vector is constant for all wave numbers on the circle, and of course, so is the frequency, so that for all waves with waves at that frequency, the magnitude of the energy flux is constant as long as the amplitude is the same for the waves. The diagram is very helpful in visualizing the relation between wave number vector and group velocity, and it is immediately apparent that that relation is not a simple one. The group velocity is neither perpendicular nor parallel to $\vec{K}$, and indeed in some cases, it is nearly in the opposite direction. This is particularly helpful in visualizing the process of reflection.

The Reflection of Rossby Waves

Consider a straight western boundary of a basin sloping at an angle θ with respect to the x-axis as shown in Fig. 16.2. Suppose a Rossby wave in the form of a packet of a beam of energy impinges on the wall from the right (the east).

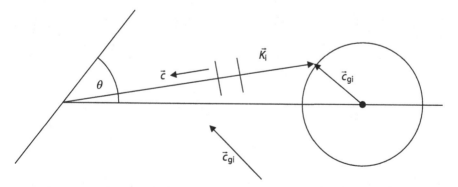

Fig. 16.2. The reflection of Rossby waves from a western wall oriented at an angle θ with respect to the x-axis. The energy propagation diagram is shown adjacent

In order for the wave energy to be moving eastward and northward as in the figure, the incident wave vector $\vec{K}_i$ must lie on the segment of the circle nearest the origin. That is, the wave must be a relatively *long* wave. For each frequency and y wave number, l, there are two choices of k, determined by the dispersion relation:

$$k = \frac{\beta}{2(-\omega)} \pm \sqrt{\frac{\beta^2}{4\omega^2} - a^2} \tag{16.22}$$

The root with the plus sign corresponds to shorter waves and a larger k and hence with group velocities to the east, while the root with the minus sign corresponds to a group velocity directed westward and is the root that must be chosen to represent the incident wave. We represent the incident wave as

$$\psi_i = A_i e^{i(k_i x + l_i y - \omega_i t)} \tag{16.33a}$$

and the reflected wave has the form

$$\psi_r = A_r e^{i(k_r x + l_r y - \omega_r t)} \tag{16.33b}$$

and during the time of interaction with the wall, the total stream function for our *linear* problem is the simple sum of the two waves:

$$\psi = \psi_i + \psi_r \tag{16.34}$$

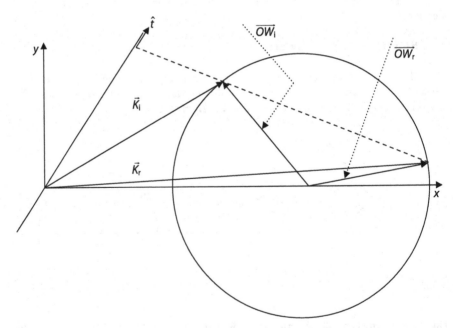

Fig. 16.3. The incident and reflected wave numbers and their position on the wave propagation circle

Let $\hat{t}$ be the *tangent* vector to the boundary (see Fig. 16.3):

$$\hat{t} = \hat{x}\cos\theta + \hat{y}\sin\theta$$

On the boundary, x and y are related by $y = x\tan\theta$. Thus, on the wall where the total stream function must be a constant (and we may choose the constant to be zero),

$$\psi = 0 = A_i e^{i[(k_i + l_i\tan\theta)x - \omega_i t]} + A_r e^{i[(k_r + l_r\tan\theta)x - \omega_r t]} \tag{16.35}$$

For this to be true for *all* x along the wall and for *all* time, it is necessary that

$$\omega_r = \omega_i \tag{16.36a}$$

$$k_r + l_r\tan\theta = k_i + l_i\tan\theta \tag{16.36b}$$

$$\Rightarrow A_r = -A_i \tag{16.36c}$$

The first condition that the frequency be conserved under reflection (which we have seen before in our study of internal waves) means that both the incident and reflected wave *must lie on the same wave number circle*. The second condition requires that the component tangent to the wall of both the incident and reflected waves must be equal. These, plus the *radiation condition* that the reflected wave has its energy directed away from the wall is sufficient to determine the position of the reflected wave on the wave number circle. Note that the magnitude of the wave number is *not conserved under reflection*. Indeed, the wave number vector is lengthened in the reflection process, i.e., *the reflected wave has a shorter wave length than the incoming wave*. Note that since the amplitude is conserved, the energy of the reflected wave per unit horizontal area is *larger than the energy of the incoming wave*. Yet energy must be conserved. Since both the incoming and outgoing waves are on the same wave number circle and the amplitude of the wave is preserved, the magnitude of the vector $\overrightarrow{OW}$ is preserved under reflection. Thus, the energy *flux* of the reflected and incident waves must be the same. High-energy, slow-moving wave packets leaving the wall are balanced by relatively low energy, rapidly moving packets impinging on the wall. It is left for the student to verify that the group speeds meet that condition. It is also left as an exercise for the reader to show that the angle of incidence of the group velocity is equal to the angle of reflection of the reflected wave packet *with respect to the boundary*. That is, the reflection process is specular.

In the special case when $\theta = \pi/2$, i.e., a wall along a longitude, the y wave number is conserved under reflection and the incident and reflected x-wave numbers satisfy

$$k_i = \frac{\beta}{(-2\omega)} - \sqrt{\frac{\beta^2}{4\omega^2} - a^2}$$

$$\tag{16.37}$$

$$k_r = \frac{\beta}{(-2\omega)} + \sqrt{\frac{\beta^2}{4\omega^2} - a^2}$$

For very low frequency waves, the discrepancy for the x wave number will be very great, and the western boundary of an ocean acts then as a source of very short (in x) scale of energy. The reflected wave will have the same zonal velocity as the incident wave, but its meridional velocity will be much larger.

The group velocity in the x-direction in the limit of very short x wave number will be of the order

$$c_{gx} = O\left(\frac{\beta}{k^2}\right)$$

and is directed eastward. In the presence of a large-scale (Sverdrup dynamics) zonal current drift, U, the net group velocity will be

$$c_{gxnet} = U + \frac{\beta}{k^2}$$

If U were negative, all scales with k larger than $\sqrt{\beta/U}$ would not escape from the generation region. This gives, as a characteristic length scale for a zone of high wave number energy near the western boundary, $\delta_I = \sqrt{U/\beta}$, which is the characteristic scale of the western boundary current in the inertial theory of the Gulf Stream. When the large-scale flow is directed *eastward*, U is positive and the energy is not trapped. This corresponds to the fact that purely inertial models for the Gulf Stream fail in regions of eastward Sverdrup flow. It is left to the student to calculate the characteristic distance over which the eastward propagating energy, when $U < 0$, would decay in the presence of lateral friction (by calculating the diffusion time for wave number k and using the group velocity) to deduce the scale of frictional models of the Gulf Stream. This is a good example of how an understanding of fundamental wave dynamics can give us insight into even the problems of steady circulation theory.

The Spin-Down of Rossby Waves

In regions far from lateral boundaries, the principal dissipative agent is bottom friction. It is beyond the scope of this course to review in a complete fashion the nature of the viscous boundary layer, the Ekman layer, and the student can refer to several texts (e.g., Greenspan 1968 or Pedlosky 1987) for a full discussion. Physically, and for simplicity let's think of a flat bottom there will exist a region of the order of $\delta_E = (A_v/f_0)^{1/2}$ near the bottom boundary where the vertical shear of the velocity will be strong so as to allow the fluid to satisfy the no-slip condition at the bottom. The thickness of this region, δ_E, depends on the Coriolis parameter and the coefficient of vertical mixing of momentum A_v, and it is generally very thin. In that layer, the fluid dynamics is no longer geostrophic, and the presence of friction allows fluid to flow across lines of constant pressure from high pressure to low pressure. Under a region of cyclonic vorticity where there will be a low pressure center, the flow in the boundary layer will converge towards the cyclone's center. Since the flow is incompressible, that lateral convergence must lead to a vertical flux of fluid out of the boundary layer into the geostrophic region above. That vertical velocity is given by the relation:

$$w(x,y,0)=w_{\mathrm{E}}=\left(\frac{A_{\mathrm{v}}}{2f_0}\right)^{1/2}\zeta \tag{16.38}$$

where ζ is the vorticity of the geostrophic flow. Note that since ζ is of the order U/L, the vertical velocity satisfies our general scaling expectation between w and the horizontal velocity, i.e., that $w = O(Ud/L)$ where d is the vertical scale of the motion. In the boundary layer, d is $\delta_{\mathrm{E}} = (A_{\mathrm{v}}/f_0)^{1/2}$, and so the result is certainly plausible. The student is encouraged to examine the cited reference for details.

For the purposes of the wave problem in quasi-geostrophic flow, the effect is to alter the equation for mass conservation for the *layer of geostrophic flow*. That comprises most of the layer (see Fig. 16.4) except for the boundary layer.

Redoing the vertical integral of the continuity equation now yields

$$
\begin{aligned}
D(u_x+v_y) &=-w(\text{top})+w(\text{bottom}) \\
&=-\frac{d\eta}{dt}+w_{\mathrm{E}} \\
&=-\frac{d\eta}{dt}+\left(\frac{A_{\mathrm{v}}}{2f_0}\right)^{1/2}\zeta
\end{aligned} \tag{16.39}
$$

Redoing the steps leading to the potential vorticity equation then yields an extra term on the right-hand side of the vorticity equation and the potential vorticity equation such that now we have

$$
\begin{aligned}
\frac{dq}{dt}\approx\frac{d}{dt}\left[\nabla^2\psi-\frac{1}{L_{\mathrm{D}}^{2}}\psi+\frac{f_0h_{\mathrm{b}}}{D_0}+\beta y\right] &=-\frac{f_0}{D_0}\left(\frac{A_{\mathrm{v}}}{2f_0}\right)^{1/2}\nabla^2\psi \\
&=-\frac{1}{T_{\mathrm{s}}}\nabla^2\psi
\end{aligned} \tag{16.40}
$$

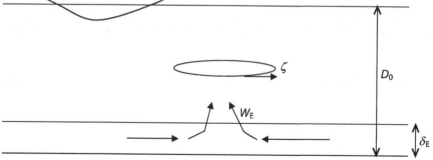

Fig. 16.4. The genesis of vertical motion pumped out of the bottom Ekman layer by cyclonic geostrophic motion above the bottom

where T_S is a characteristic decay time for the system due to bottom friction:

$$T_s = \frac{D_0}{(A_v f_0 / 2)^{1/2}}$$
(16.41)

Note that this time, called the *spin-down time* (or spin-up time for optimists), increases with the depth and decreases as the mixing coefficient and the rotation get larger.

For linear Rossby waves, the wave equation becomes

$$\frac{\partial}{\partial t}[\nabla^2 \psi - a^2 \psi] + \beta \psi_x = -\frac{1}{T_s} \nabla^2 \psi$$
(16.42)

Solutions for plane waves can be found in the form:

$$\psi = A e^{-\sigma t} e^{i(kx+ly-\omega t)}$$
(16.43)

where σ is the frictional decay rate. Inserting the above form in the wave equation yields, after equating real and imaginary parts of the dispersion relation,

$$\sigma = \frac{K^2}{K^2 + a^2}\left(\frac{1}{T_s}\right)$$
(16.44a)

$$\omega = -\beta \frac{k}{K^2 + a^2}$$
(16.44b)

The frequency wave number relation is unchanged, and the decay rate is in fact one over the spin-down time slightly modified by the scale. Note that when the horizontal scale is very large compared to the deformation radius, $K \ll a$, the decay rate is small, while for short length scales for which the above equality is reversed, i.e., when the scale is small compared to a deformation radius, the decay rate becomes *independent of scale* (This would exactly occur if there were an upper rigid lid instead of a free surface. Why?).

For our previous work on waves to have relevance it is necessary that we can observe at least several oscillations before the wave decays. That is the basis of our approximation that inviscid theory is pertinent to the wave problem. So we have been implicitly assuming all along that

$$\omega T_s = 2\pi \frac{T_s}{T\text{period}} \gg 1$$
(16.45)

Laplace Tidal Equations
and the Vertical Structure Equation

Let's return to the linearized wave equations before the gravity waves are filtered out by the quasi-geostrophic approximation. What we will see now is that the analysis of the homogeneous model can be carried over, in important cases, to the motion of a stratified fluid. A vertical modal decomposition can be done for these cases, and we will be able to show that the equations for *each vertical mode* are analogous to the equations for the single layer. Exactly what that relationship is will be the subject of our development that follows.

To keep the discussion simple, we will consider *hydrostatic motion* but not necessarily geostrophic motion. We will also relax the β-plane approximation and consider linearized motion on the sphere. Our coordinates will be θ for latitude, ϕ for longitude, and z for the elevation above the Earth's spherical surface as seen in Fig. 17.1. The velocities in the zonal, meridional and vertical directions will be u, v and w. As before, we will separate the pressure and density into the values those variables have in the rest state plus a small perturbation due to the motion

$$p_{total} = p_0(z) + p(\phi,\theta,z,t) \tag{17.1a}$$

$$\rho_{total} = \rho_0(z) + \rho(\phi,\theta,z,t) \tag{17.1b}$$

and we will assume that the density field of the basic state ρ_0 changes only slightly over the vertical extent of the fluid (for a compressible fluid like the atmosphere, see Andrews et. al. 1987), i.e.,

$$\frac{D\partial\rho_0}{\rho_0\partial z} \ll 1$$

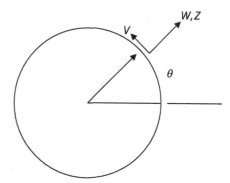

Fig. 17.1.
The coordinate system for the hydrostatic
equations of motion on the sphere

For linearized, inviscid motion, the equations for the perturbations become

$$\rho_0\left[u_t - \overbrace{2\Omega\sin\theta v}^{a} + \overbrace{2\Omega\cos\theta w}^{b}\right] = -\frac{p_\phi}{R\cos\theta} \tag{17.2a}$$

$$\rho_0\left[v_t + 2\Omega\sin\theta u\ \right] = -\frac{p_\theta}{R} \tag{17.2b}$$

$$-\overbrace{\rho_0 2\Omega\cos\theta u}^{c} = -\overbrace{p_z}^{d} - g\rho \tag{17.2c}$$

$$\frac{u_\phi}{R\cos\theta} + \frac{(v\cos\theta)_\theta}{R\cos\theta} + w_z = 0 \tag{17.2d}$$

$$\rho_t + w\frac{\partial\rho_0}{\partial z} = 0 \tag{17.2e}$$

We have used subscripts for differentiation. R is the (constant) Earth's radius. We have also assumed in the last equation that the motion is *adiabatic*. In the momentum equations, we have included each of the components of the Coriolis acceleration, $2\vec{\Omega}\times\vec{u}$, and please note that while the contribution in Eq. 17.2a of the component of the Earth's rotation that is tangent to the Earth's surface, $2\Omega\cos\theta$, involves the weak vertical velocity (we are assuming the vertical scale of the motion is much less than its horizontal scale), its contribution in the vertical equation of motion depends on the much stronger zonal velocity. The issue here is if we ignore this contribution in the zonal momentum equation, can we also consistently ignore it in the vertical equation of motion? It is often said that if one approximation is made, the other *must* be made; otherwise if term (b) is absent but term (c) retained, the dot product of the velocity with the momentum equations would have the Coriolis force doing work on the fluid, an obvious absurdity since it is always perpendicular to the velocity. But saving us from absurdity is not a justification for an approximation. We must show term (c) is small if term (b) is. Moreover, the smallness of term (c) *in its equation* must be measured by the same parameter of smallness as term (b) is measured in its equation. If there were different parameters that measured the relative smallness of those terms in each of their equations, we might find a situation where one parameter was small and the other $O(1)$. So, we must see whether a sensibility scaling argument will let us *always* ignore both terms simultaneously.

The ratio of term (b) to term (a) is obviously of the order of the vertical to horizontal velocity. If the scale of the former is W and that of the latter is U, we know from the continuity equation that

$$\frac{W}{U} = O\left(\frac{D}{L}\right) \ll 1$$

where D and L are the vertical and horizontal scales of motion. Thus if $\delta = D/L \ll 1$, we can ignore term (b) compared to term (a) in the zonal momentum equation. If the Coriolis acceleration enters at the lowest order into the dynamics, then this tells us that the *scale* of the pressure field, P, must be $P = \rho_0 2\Omega UL$.

This does not imply that there is a balance between the horizontal pressure gradient and the Coriolis acceleration, only that they are both $O(1)$ terms in the momentum equation. With that scaling for the pressure, the vertical pressure gradient will be of the order

$$p_z = O(\rho_0 2\Omega UL/D) \tag{17.3}$$

so that the ratio of term (c) to term (d) will be

$$\frac{\rho_0 2\Omega u \cos\theta}{p_z} = O(D/L) \tag{17.4}$$

Thus, if $D/L \ll 1$, the horizontal component of the Earth's rotation can be ignored in both the zonal and vertical momentum equations. Our final set of approximate equations is therefore

$$\rho_0[u_t - 2\Omega \sin\theta v] = -\frac{p_\phi}{R\cos\theta} \tag{17.5a}$$

$$\rho_0[v_t + 2\Omega \sin\theta u] = -\frac{p_\theta}{R} \tag{17.5b}$$

$$0 = -p_z - g\rho \tag{17.5c}$$

$$\frac{u_\phi}{R\cos\theta} + \frac{(v\cos\theta)_\theta}{R\cos\theta} + w_z = 0 \tag{17.5d}$$

$$\rho_t + w\frac{\partial \rho_0}{\partial z} = 0 \tag{17.5e}$$

As before, we can write the last equation as

$$\rho_t - w\rho_0 N^2/g = 0, \quad N^2 \equiv -\frac{g}{\rho_0}\frac{d\rho_0}{dz} \tag{17.6}$$

and the use of the hydrostatic equation allows us to eliminate ρ completely from the equations by writing the adiabatic equation as

$$p_{zt} + \rho_0 w N^2 = 0 \tag{17.7}$$

We will consider only those situations in which the bottom of the fluid is flat at $z = -D$ and the top consists of a free surface. When the bottom is flat, we can separate the variables in the problem into a function of z and a function of horizontal and time variables. Following the treatment in Moore and Philander (1977) and Pedlosky (1987), we write,

$$\begin{pmatrix} u \\ v \end{pmatrix} = \begin{bmatrix} U(\phi,\theta,t) \\ V(\phi,\theta,t) \end{bmatrix} F(z) \tag{17.8a}$$

$$w = W(\phi,\theta,t)G(z) \tag{17.8b}$$

$$\frac{p}{\rho_0} = g\eta(\phi,\theta,t)F(z) \tag{17.8c}$$

Note that U should not be confused with our earlier use of the same symbol for the horizontal velocity scale. It is now a variable and a function of horizontal coordinates and time. The functions F and G are functions only of z, and they must be determined by the physics. Our first goal is to derive the governing structure equations for those functions. Finally, the variable η is a function that represents the horizontal structure of the pressure field. At this stage, it has nothing at all to do with the actual elevation of the fluid's free surface.

If we insert these forms in the equations of motion, we find first from the horizontal momentum equations, since $F(z)$ is a common factor,

$$U_t - fV = -\frac{g\eta_\phi}{R\cos\theta} \tag{17.9a}$$

$$V_t + fU = -\frac{g\eta_\theta}{R} \tag{17.9b}$$

Aside from the use of spherical coordinates to represent the horizontal pressure gradient, these are exactly the linearized momentum equations for a single layer of fluid whose velocity components are U and V and whose free surface elevation is η.

The same process for the continuity equation is not so simple: w depends on the function G while u and v are proportional to F. This leads to

$$\frac{U_\phi}{R\cos\theta} + \frac{(V\cos\theta)_\theta}{R\cos\theta} + W\frac{G_z}{F} = 0 \tag{17.10}$$

All terms except the ratio of G_z/F are independent of z, while each term in this ratio is a function only of z. The only way this can be consistent (this is familiar from the standard separation of variables) is if the ratio is a *constant*. We define the constant as

$$\frac{G_z}{F} = \text{constant} = \frac{1}{h} \tag{17.11}$$

The constant h is called the *equivalent depth* (we shall see why shortly) but at this stage of our analysis, it is only a separation constant. This allows us to write the continuity equation in the suggestive form:

$$\frac{U_\phi}{R\cos\theta} + \frac{(V\cos\theta)_\theta}{R\cos\theta} + \frac{W}{h} = 0 \tag{17.12}$$

Inserting the forms into the adiabatic equation in its form involving the vertical pressure gradient Eq. 17.7 yields

$$\eta_t + W\frac{G}{F_z}\frac{N^2}{g} = 0$$

$$\Rightarrow \eta_t + W\frac{G}{G_{zz}}\frac{N^2}{gh} = 0 \tag{17.13}$$

and it is clear, as in our discussion of the continuity equation, that the coefficient of the term W in the last equation must be a constant. We choose that constant to be -1. There is no loss of generality in doing this. Choosing any other constant would only alter the definition of h. The properly skeptical student should try it.

With this choice, the adiabatic equation becomes

$$\eta_t = W \tag{17.14}$$

which is *not* a boundary condition but is rather the separated form of the adiabatic equation, although the form is delightfully suggestive of the boundary condition for a single homogeneous layer. With the above choices for the separation constant, the function G now satisfies (and it is here that it would be clear that any other choice than -1 would only alter the definition of h)

$$G_{zz} + \frac{N^2}{gh}G = 0 \tag{17.15}$$

This is a homogeneous differential equation with, generally, nonconstant coefficients, since N is a function of z and with a free parameter h. The problem is not complete until the boundary conditions are established.

In order to have w vanish on $z = -D$, we must take

$$G(z) = 0, \quad z = -D \tag{17.16}$$

At the free surface, the conditions are that the free surface displacement, which here we will call z_T, satisfies

$$w = WG(z_T)\frac{\partial z_T}{\partial t} \tag{17.17}$$

while the total pressure is atmospheric pressure, which we will take to be a constant (zero), thus

$$\begin{aligned}
p_{\text{total}} &= p_0(z_T) + g\eta F(z_T) \\
&\approx p_0(0) + \frac{dp_0}{dz}z_T + \ldots + g\eta F(0)
\end{aligned} \tag{17.18}$$

keeping only linear terms.

A time derivative of the last equation combined with the kinematic condition then yields

$$gWG(0) = g\eta_t F(0) \tag{17.19}$$

but from the continuity equation, this implies that $z = 0$:

$$G(0) = F(0) = hG_z(0) \Rightarrow G_z - G/h = 0, \quad z = 0 \tag{17.20}$$

which is the final condition for G. We then have an eigenvalue problem for the function $G(z)$, whose eigenvalues are h. Note that the problem could also be written in terms of $F(z)$.

Using the above relations between F and G, we obtain as an equally valid alternative problem:

$$\left(\frac{F_z}{N^2}\right)_z + \frac{1}{gh}F = 0 \tag{17.21a}$$

$$F_z = 0, \quad z = -D \tag{17.21b}$$

$$F_z + \frac{N^2}{g}F = 0, \quad z = 0 \tag{17.21c}$$

The advantage of the second formulation is that the eigenvalue h is not in the boundary condition.

The equations for either G or F can be solved numerically, and the eigenvalue is found along with the structure of the solution in z. Insight into the nature of the problem can be gained by examining the case for the constant N.

In that case, the solution for $G(z)$, which satisfies the boundary condition at $z = -D$, is

$$G = A\sin m(z+D), \quad m^2 \equiv \frac{N^2}{gh} \tag{17.22}$$

m is the vertical wave number of the solution (it will be quantized since the region is finite), and

$$\lambda_z = \frac{2\pi}{m} \tag{17.23}$$

will be the vertical scale of the motion in the mode that has that value of m as the vertical wave number. Note that the vertical scale of the motion *is not* h. Indeed, if we define the *vertical scale height for the density*

$$h_\rho = -\frac{\rho_0}{d\rho_0/dz} \tag{17.24}$$

as the scale over which the density changes by its own magnitude (this is much greater than the depth of the ocean for realistic oceanic density gradients), the vertical scale of the *motion* is given by

$$\lambda_z = 2\pi\sqrt{hh_\rho} \tag{17.25}$$

so that the vertical scale of the motion is essentially the geometric *mean* of the equivalent depth and the density scale height.

The eigenvalue relation for h is obtained from the final boundary condition at $z = 0$ and yields

$$m \cos mD - \frac{1}{h} \sin mD = 0 \tag{17.26a}$$

$$\tan(mD) = mh = \frac{N^2}{gm} \tag{17.26b}$$

or

$$\tan(mD) = \left(\frac{N^2 D}{g} \right) \frac{1}{mD} \tag{17.27}$$

We note as we have before that

$$N^2 D / g = -\frac{D}{\rho_0} \frac{d\rho_0}{dz} = \frac{D}{h_\rho} \approx \frac{\Delta\rho_0}{\rho_0} \ll 1 \tag{17.28}$$

Thus the roots of the dispersion relation split into two classes. The first class has roots for which mD is $O(1)$. In that case, the right-hand side of the dispersion relation is essentially zero, and the solutions correspond to the zeros of the tangent function, or

$$mD = j\pi, \quad j = 1,2,3,\dots \tag{17.29}$$

There are an infinite number of such roots corresponding to

$$m = \frac{j\pi}{D} \tag{17.30}$$

and since $m^2 = N^2 / gh$, the associated equivalent depth for mode j is

$$gh_j = \frac{N^2 D^2}{j^2 \pi^2} \tag{17.31}$$

Note that for this mode, the horizontal equations will contain an equivalent long wave gravity wave speed:

$$c_j = \sqrt{gh_j} = \frac{ND}{j\pi} \ll \sqrt{gD} \tag{17.32}$$

These equivalent speeds are the long wave speeds for *internal gravity modes* of vertical mode number j and are much slower than the homogeneous phase speed for long waves $\sqrt{gD}$.

The modal structures for each j are simply

$$G_j = \sin(j\pi z / D), \quad j = 1,2,3\dots \tag{17.33a}$$

$$F_j = \cos(j\pi z / D), \quad j = 1,2,3 \tag{17.33b}$$

and although these form a complete set for the representation of w (the sine series is complete), it is clear that the set of functions F_j, while containing an infinite number of functions, which are all orthogonal, is not complete, since the cosine series lacks the constant term. In other words, we have not found the barotropic mode that contains zero vertical velocity.

We must reexamine the dispersion relation. We previously assumed that mD was $O(1)$. That may not always be the case. Indeed, as $mD \longrightarrow 0$, the dispersion relation becomes

$$\tan mD \approx mD = \frac{N^2 D}{gmD}$$

$$\Rightarrow m^2 D^2 = \frac{N^2 D}{g} \tag{17.34}$$

but by definition,

$$m^2 = \frac{N^2}{gh} \tag{17.35}$$

or for this mode subscripted zero,

$$h_0 = D \tag{17.36}$$

or

$$m_0 = \sqrt{\frac{N^2}{gD}}$$

and so

$$m_0 D = \sqrt{\frac{N^2 D^2}{gD}} \ll 1 \tag{17.37}$$

so that the function

$$F_0 = \cos m_0(z + D) \tag{17.38}$$

hardly varies at all in z, i.e., the function is very nearly z-independent. This is the barotropic mode.

To sum up, for linear, inviscid motion of a stratified fluid on the sphere, when the fluid has a flat bottom, we can separate the motion into an infinite number of vertical modes. Each mode satisfies a set of equations for its horizontal structure, which is *identical to that of a homogeneous layer of fluid possessing a long gravity wave speed* $c_j = (gh_j)^{1/2}$. That is, it behaves as a homogeneous layer with the equivalent depth h_j, which itself is one of the eigenvalues of the vertical structure equation. This is the only way stratification enters the problem, i.e., by determining the equivalent depths and phase speeds and by determining the vertical structure of the modes. Note that for each mode in z, the vertical structure is maintained as that mode propagates, reflects or dissipates as time progresses. Thus, *all our previous work on the dynamics of Poincaré, Kelvin and Rossby waves for a homogeneous layer can be carried over, mode by mode, to a stratified layer as long as the motion is hydrostatic and the bottom is flat.* If the bottom is not flat, it is not possible to separate the motion into individual modes. The topography will mix the modes together, and the modal description is no longer useful. The equations for the horizontal structure are called *Laplace's tidal equations*, because in their original application to a homogeneous fluid, they are with appropriate added forcing terms the equations for the ocean tides. To recapitulate those equations we have

$$U_t - fV = -\frac{g\eta_\phi}{R\cos\theta} \tag{17.39a}$$

$$V_t + fU = -\frac{g\eta_\theta}{R} \tag{17.39b}$$

$$\frac{U_\phi}{R\cos\theta} + \frac{(V\cos\theta)_\theta}{R\cos\theta} + \frac{W}{h} = 0 \quad \text{and} \tag{17.39c}$$

$$\eta_t = W \tag{17.39d}$$

Note that if the last two equations are combined,

$$\eta_t + h\left[\frac{U_\phi}{R\cos\theta} + \frac{(V\cos\theta)_\theta}{R\cos\theta}\right] = 0 \tag{17.40}$$

and the correspondence to the dynamics of a shallow layer of fluid of depth h is complete.

Equatorial Beta-Plane and Equatorial Waves

The equator is a special region dynamically, most obviously because there the vertical component of the Earth's rotation vanishes. It turns out to be, in consequence, a region in which certain linear waves have unusually strong signals and are involved in some important atmospheric and oceanic phenomena such as the Quasi-Biennial Oscillation in the atmosphere and the El Niño (ENSO) phenomenon in the ocean (and atmosphere). Good, useful references that describe in detail those phenomena are Andrews et al. (1987) for the former and Philander (1990) for the latter.

To see intuitively why the equator might be such as special zone, consider heuristically a Poincaré wave packet near the equator with frequency

$$\omega = \left(f^2 + c_0^2 (k^2 + l^2) \right)^{1/2} \tag{18.1}$$

and we note that near the equator where f vanishes, the y-dependence of the Coriolis parameter cannot be neglected. The dispersion relation is of the class of relations discussed in our first lecture where the relation between frequency and wave number also *explicitly* includes a dependence on a spatial variable, in this case latitude or locally, y, i.e.,

$$\omega = \Omega(\vec{K}, f(y)) \tag{18.2}$$

As we noted in the first lecture, the wave vector for a slowly varying packet satisfies

$$\frac{d\vec{K}}{dt} = -\nabla \Omega = -\beta \frac{f}{\omega} \hat{y} \tag{18.3}$$

where $\hat{y}$ is a unit vector in the meridional direction. Remember that the gradient on the right-hand side of the equation for the rate of change of the wave vector is the gradient with respect to the explicit dependence of the dispersion relation on spatial variables, in this case only y.

That means that as the packet propagates, the frequency and x-wave number, k, will be constant in the packet; only l will change. The dispersion relation for Poincaré waves implies that as the packet moves to higher latitudes where f^2 is larger, the y-wave number must decrease to keep the frequency constant. Finally at the latitude y_c such that

$$f(y_c) = \left[\omega^2 - c_0^2 k^2 \right]^{1/2} \tag{18.4}$$

the y wave number vanishes. Beyond that point, l becomes *imaginary*, leading to spatial decay with y for $y > y_c$. This produces a trapping zone around the equator in which the wave energy, which would normally disperse in two dimensions, is trapped within a wave guide as a consequence of the increase with latitude of the square of the Coriolis parameter, an effect that is clearly symmetric about the equator. Similarly, for Rossby waves where

$$\omega = -\beta \frac{k}{k^2 + l^2 + f^2/c_0^2} \tag{18.5}$$

the same trapping effect must occur. Note that for both Poincaré and Rossby waves, the meridional component of the group velocity vanishes when the y wave number vanishes so that the wave energy will not cross the critical latitude and will be reflected back into the *equatorial wave guide*. Also note that as the Coriolis parameter vanishes, the minimum frequency of Poincaré waves approaches the maximum frequency of Rossby waves, and so the two wave types cannot be expected to be as well-separated in the frequency domain as they are in mid-latitudes.

Thus, overall, the equatorial band will act as a wave guide for both gravity and Rossby waves. We expect the wave modes to be trapped meridionally and the propagation to be basically along the equator. This means the waves will generally not disperse their energy over more than the zonal direction, and consequently the amplitude of the waves and their influence can be anticipated to be more important for equatorial dynamics in general than in mid-latitudes. It remains for us to move beyond this heuristic discussion to find the nature of the waves in the equatorial zone.

The Equatorial Beta-Plane

We will assume that the wave motions have a large enough horizontal scale to ensure that the wave motion is hydrostatic. We will also only consider cases in which the ocean bottom is considered flat, and in fact, we will not consider any interaction with the bottom. In that case, as we saw in the last lecture, we can resolve the wave motion on a set of vertical normal modes, each mode yielding an equivalent depth h_n and a corresponding long wave speed c_n, both of which come from the eigenvalue problem described in the previous lecture.

For the linear inviscid problem, the equations of motion are

$$\frac{\partial U_n}{\partial t} - fV_n = -\frac{g}{R\cos\theta}\frac{\partial \eta_n}{\partial \phi} \tag{18.6a}$$

$$\frac{\partial V_n}{\partial t} - fU_n = -\frac{g}{R}\frac{\partial \eta_n}{\partial \theta} \tag{18.6b}$$

$$\frac{\partial \eta_n}{\partial t} + h_n \left[\frac{1}{R\cos\theta}\frac{\partial U_n}{\partial \phi} + \frac{1}{R\cos\theta}\frac{\partial}{\partial \theta}(V_n \cos\theta) \right] = 0 \tag{18.6c}$$

If the motion is limited to a narrow region, L, around the equator such that $L/R \ll 1$, we can expand the trigonometric functions in the above equations, i.e.,

$$f = \underbrace{f(0)}_{=0} + \beta y + \ldots = \beta y + O(L/R)^3 \tag{18.7a}$$

$$\cos\theta = 1 + O(L/R)^2 \tag{18.7b}$$

This allows us to define the local Cartesian coordinate system:

$$x = R\cos\theta[\phi - \phi_0] = R(\phi - \phi_0) + O(L/R)^2 \tag{18.8a}$$

$$y = R\theta + O(L/R)^2 \tag{18.8b}$$

In these terms, the equations become the simpler set, valid on the *equatorial beta-plane*:

$$U_{n_t} - \beta y V_n = -g\eta_{n_x} \tag{18.9a}$$

$$V_{n_t} + \beta y U_n = -g\eta_{n_y} \tag{18.9b}$$

$$\eta_{n_t} + h_n[U_{n_x} + V_{n_y}] = 0 \tag{18.9c}$$

Of course, we must check after the fact that our solution does satisfy the condition of being localized in the vicinity of the equator. It is also easy to add forcing terms to each of the momentum equations to represent the action of a wind stress, and the exercise is left to the student to trace the development of the equations with such forcing terms present.

Now the heuristic discussion at the start of the lecture leads us to anticipate that the wave modes will be contained in a wave guide, a sort of naturally produced equatorial channel. In that case, we might anticipate that the modes will be analogous to the modes we found in the channel problem for mid-latitudes. In that case, we found Poincaré, Kelvin and Rossby modes. It was a straightforward business in the mid-latitude case to write the problem in terms of the free surface height. With the strong variation of f in the equatorial case, it turns out be far simpler to pose the problem in terms of the meridional velocity (we noted in the mid-latitude case that the eigenstructure for the meridional velocity was far simpler than for either the free surface perturbation or the zonal velocity). However, based on our experience in the mid-latitude channel, we also might anticipate that we should be alert to a wave mode for which the meridional velocity is identically zero, for that is one of the chief characteristics of the Kelvin wave. Hence, before we formulate the wave problem in terms of the meridional velocity, we should check to see whether a mode exists in which V_n is identically zero. If that were so, we would have

$$U_t = -g\eta_x \tag{18.10a}$$

$$\beta y U = -g\eta_y \tag{18.10b}$$

$$\eta_t + hU_x = 0 \tag{18.10c}$$

We have suppressed the explicit subscript notation, and the student is expected to realize that the following development is pertinent to each mode n, each with its own equivalent depth h.

Eliminating the free surface elevation between the first and third equations yields

$$U_{tt} - ghU_{xx} = 0 \qquad (18.11)$$

which is again the classical one-dimensional wave equation whose general solution is

$$U^{\pm} = U^{\pm}(x \pm ct, y) \qquad (18.12)$$

One solution, $U^+ = U^+(x + ct, y)$, propagates to the west with no change of shape, while the other solution, $U^- = U^-(x - ct, y)$, propagates eastward with no change of shape. These are reminiscent of the Kelvin waves in a channel. In that case, the right-moving wave "leaned" against the lower wall, and the left-moving wave "leaned" against the upper wall. In the present case, there are *no* walls, only the equator itself, and we have to check whether either of these solutions has a y-structure consistent with equatorial trapping of the disturbance. The y-derivative of the x-momentum equation and the x-derivative of y-momentum equation to eliminate the free surface term yields

$$U_{yt} = \beta y U_x \qquad (18.13)$$

Now for each possible solution we have

$$U_t^{\pm} = \pm c U_t^{\pm} \qquad (18.14)$$

which when inserted in $U_{yt} = \beta y U_x$ yields

$$\frac{\partial}{\partial y}\left(U_x^{\pm}\right) \mp \frac{\beta y}{c}\left(U_x^{\pm}\right) = 0 \quad \text{and}$$

$$\Rightarrow U_x^{\pm} = e^{\pm(\beta y^2 / c)} F^{\pm}(x \pm ct) \qquad (18.15)$$

where F is an arbitrary function.

Thus, if the region is unbounded in y only, the (−) solution is acceptable, since the (+) solution diverges at $|y| = \infty$. There is *only an eastward moving Kelvin wave mode*:

$$U_{\text{Kelvin}} = e^{-(\beta y^2 / c)} F(x - ct) \qquad (18.16)$$

with a corresponding free surface height,

$$\eta_{\text{Kelvin}} = \frac{c}{g} e^{-(\beta y^2 / c)} F(x - ct) \qquad (18.17)$$

The mode moves eastward with the long wave speed. For each vertical mode n, that wave speed is $c = c_n = \sqrt{gh_n}$. The decay scale is

$$L_{\text{eq}} = \left(\frac{c_n}{\beta}\right)^{1/2}$$

which is *the equatorial deformation radius*.

Table 18.1.
Quantities for the first five baroclinic modes (Moore and Philander 1977)

n	h_n (cm)	c_n (cm s^{-1})	L_{eq} (km)	T (days)
1	60	240	325	1.5
2	20	140	247	2.0
3	8	88	177	2.6
4	4	63	165	3.1
5	2	44	138	3.6

To get a feeling for the quantities involved, Table 18.1 (from the article by Moore and Philander 1977) gives the pertinent quantities for the first five vertical *baroclinic* modes. The quantities refer to the Equatorial Atlantic but are typical.

The appropriate time scale is determined by the relation

$$\frac{L_{eq}}{T} = c_n, \quad \Rightarrow T = \frac{1}{\sqrt{\beta c_n}}$$

Note that the equatorial deformation radius depends on the square root of the long wave speed, not on the speed itself as in mid-latitudes. In mid-latitudes, the deformation radius is

$$L_D = \frac{c_n}{f} \tag{18.18}$$

but at the equator $f = \beta y$. If we set y to be of the order of the equatorial scale, this yields

$$L_{eq} = \frac{c_n}{\beta L_{eq}} \tag{18.19}$$

whose solution yields our previous definition. Note that the time scale above satisfies

$$T = \beta L_{eq} \tag{18.20}$$

as in Rossby waves.

For the *barotropic mode*, the equivalent depth h is of the order of the fluid depth, D. This yields

$$c_0 = \sqrt{gD} = O(200) \text{ m s}^{-1} \tag{18.21a}$$

$$L_{eq} = 3000 \text{ km} \tag{18.21b}$$

so that the "trapping" scale is of the order of the planetary scale. In that case, the equatorial wave guide has little sense, since it is global and the barotropic mode must be considered separately. Fortunately, most of the equatorial response that seems to be relevant is in the baroclinic modes, and indeed the equatorial Kelvin wave has been clearly identified in the equatorial regions (Eriksen et al. 1983).

Note that the y-structure of the Kelvin mode is a Gaussian and that U is in *geostrophic balance with the pressure field*.

We now return to the task of deriving a wave equation for the rest of the wave modes. Taking a time derivative of the x-momentum equation and using the mass conservation equation yields

$$U_{tt} - \beta y V_t = -g\eta_{xt} = gh(U_{xx} + V_{xy})$$

$$\Rightarrow \left[U_{tt} - ghU_{xx} = \beta y V_t + ghV_{yx} \right] \tag{18.22}$$

while operating in the same way on the y-momentum equation yields

$$V_{tt} + \beta y U_t = -g\eta_{yt} = gh(U_{xy} + V_{yy})$$

$$\Rightarrow \left[V_{tt} - ghV_{yy} = -\beta y U_t + ghU_{xy} \right] \tag{18.23}$$

We now operate on the above equation for V (Eq. 18.23) with the operator

$$\left(\frac{\partial^2}{\partial t^2} - c^2 \frac{\partial^2}{\partial x^2} \right)$$

and use the equation for U (Eq. 18.22) to obtain

$$\left(\frac{\partial^2}{\partial t^2} - c^2 \frac{\partial^2}{\partial x^2} \right)\left(\frac{\partial^2}{\partial t^2} - c^2 \frac{\partial^2}{\partial y^2} \right)V = -\beta y \left(\frac{\partial^2}{\partial t^2} - c^2 \frac{\partial^2}{\partial x^2} \right)U$$

$$+ c^2 \frac{\partial^2}{\partial x \partial y}\left(\frac{\partial^2}{\partial t^2} - c^2 \frac{\partial^2}{\partial x^2} \right)U \tag{18.24}$$

or

$$\left(\frac{\partial^2}{\partial t^2} - c^2 \frac{\partial^2}{\partial x^2} \right)\left(\frac{\partial^2}{\partial t^2} - c^2 \frac{\partial^2}{\partial y^2} \right)V = -\beta y \frac{\partial}{\partial t}\left[\beta y \frac{\partial V}{\partial t} + c^2 \frac{\partial^2 V}{\partial x \partial y} \right]$$

$$+ c^2 \frac{\partial^2}{\partial x \partial y}\left[\beta y \frac{\partial V}{\partial t} + c^2 \frac{\partial^2 V}{\partial x \partial y} \right] \tag{18.25}$$

Carrying out the algebra implied by the above products yields the final equation for V:

$$\frac{\partial}{\partial t}\left[\frac{\partial}{\partial t}\left\{ \nabla^2 V - \frac{1}{c^2}V_{tt} - \frac{\beta^2 y^2}{c^2}V \right\} + \beta V_x \right] = 0 \tag{18.26}$$

Note that the final equation is a local conservation statement for the quantity in the square bracket. The student at this point is allowed to guess what that quantity really is and to verify the presumption. Also note the similarity of the equation to the wave equation at mid-latitude where the factor $\beta^2 y^2$ is replaced by f^2. Also be sure to recall that this equation holds for *each vertical mode*, i.e., for each n with a corresponding vertical structure function and a corresponding equivalent depth.

Let's try to find plane wave solutions in x and t of the form

$$V = A e^{i(kx-\omega t)} \psi(y) \tag{18.27}$$

where ψ satisfies the ordinary differential equation:

$$\frac{d^2\psi}{dy^2} + \left[\frac{\omega^2}{c^2} - \frac{\beta^2 y^2}{c^2} - k^2 - \beta \frac{k}{\omega} \right] \psi = 0 \tag{18.28}$$

Note that beyond a certain critical value of y, the form of the equation implies evanescent (or exponentially growing) behavior. Equatorward of that latitude, the function ψ will be oscillatory in y. We can put the equation in standard form by introducing a meridional coordinate scaled on the equatorial deformation radius:

$$y = \left(\frac{c}{\beta} \right)^{1/2} \xi \tag{18.29}$$

In terms of which,

$$\frac{d^2\psi}{d\xi^2} + \left[\frac{\omega^2}{\beta c} - k^2 \frac{c}{\beta} - \frac{ck}{\omega} - \xi^2 \right] \psi = 0 \tag{18.30}$$

This, interestingly enough, is exactly the Schroedinger equation for the quantum mechanical oscillator, and the solutions have been extensively studied. In the mathematics of special functions, this is the Hermite equation. It is well-known (see for example Schiff 1955) that the *only solution that is bounded at infinity is of the form*

$$\psi = \psi_j(\xi) = \frac{e^{-\xi^2/2}}{\sqrt{(2^j j! \pi^{1/2})}} H_j(\xi) \tag{18.31}$$

This, aside from the complicated constant in the denominator to make the functions orthonormal, is a Gaussian in the meridional coordinate multiplied by one of an infinite set of polynomials $H_j(\xi)$ called the Hermite polynomials. The orthogonality condition is

$$\int_{-\infty}^{\infty} \psi_j \psi_i \, d\xi = \delta_{ij} \tag{18.32}$$

The Hermite polynomials are generated according to the rule:

$$H_j(\xi) = (-1)^j e^{\xi^2/2} \frac{d^j}{d\xi^j} e^{-\xi^2/2} \tag{18.33}$$

The first few polynomials are

$H_0 = 1$

$H_1 = 2\xi$

$H_2 = 4\xi^2 - 2$ $\qquad\qquad\qquad\qquad\qquad\qquad\qquad$ (18.34)

$H_3 = 8\xi^3 - 12\xi$

$H_4 = 16\xi^4 - 48\xi^2 + 12$

Note that the solutions divide into odd and even functions because of the symmetry in y of the governing equation for ψ. Each of these functions, finite at infinity, satisfy

$$\frac{d^2\psi}{d\xi^2} + \left[(2j+1) - \xi^2 \right]\psi = 0 \qquad\qquad\qquad (18.35)$$

The first four of the eigenfunctions are shown In Fig. 18.1.

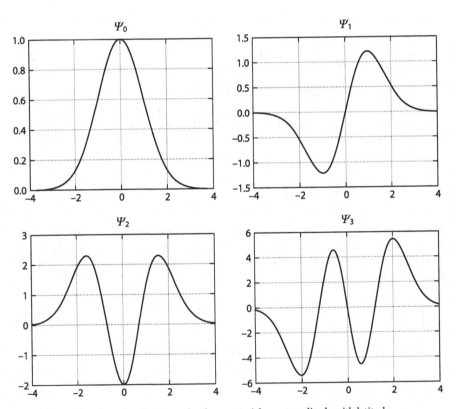

Fig. 18.1. The first four eigenfunctions for the equatorial wave amplitude with latitude

For our solution to correspond to one of these eigenfunctions and hence satisfy a condition of finiteness at infinity, we must have

$$\frac{\omega^2}{\beta c} - \frac{k^2 c}{\beta} - \frac{kc}{\omega} = 2j+1, \quad j=1,2,3,\dots \quad \text{or} \tag{18.36}$$

$$\omega^2 = c^2(k^2 + (2j+1)/L_{eq}^2) + \frac{\beta k}{\omega}c^2 \tag{18.37}$$

This should be compared to the dispersion relation at mid-latitude for Poincaré waves for a value of $f \longrightarrow 0$, while the last term on the right-hand side is similar to the effect produced by the beta term at mid-latitudes.

This is a cubic for frequency in terms of x-wave number, and it is easier to solve the quadratic for k in terms of ω. This yields the dispersion relation in the form

$$k = -\frac{\beta}{2\omega} \pm \frac{1}{2}\left[\left(\frac{\beta}{\omega} - \frac{2\omega}{c}\right)^2 - 8j\frac{\beta}{c}\right]^{1/2} \tag{18.38}$$

Before discussing the full form of this relation, it is useful to discuss limiting cases. If $\omega = O(kc)$, the last term in Eq. 18.37 would then be of $O(\beta c)$, which compared to the first term is

$$O\left(\frac{\beta}{k^2 c}\right)$$

Thus, if the gravity wave speed is much greater than the Rossby wave speed, $c > \beta / k^2$, then the last term can be neglected and we obtain the approximate dispersion relation for the Poincaré waves:

$$\omega = \pm c\left(k^2 + (2j+1)\right)^{1/2} \tag{18.39}$$

On the other hand, if ω is small, we would obtain a balance between the last term in Eq. 18.37 or the approximate equation for the equatorial Rossby mode, i.e.,

$$\omega = -\frac{\beta}{k^2 + (2j+1)/L_{eq}^2} \tag{18.40}$$

It is also easy to solve Eq. 18.38 when $j = 0$. In that case,

$$k = -\frac{\beta}{2\omega} \pm \frac{1}{2}\left(\frac{\beta}{\omega} - \frac{2\omega}{c}\right) \tag{18.41}$$

The two roots are then

$$k = -\frac{\omega}{c} \tag{18.42a}$$

$$k = -\frac{\beta}{\omega} + \frac{\omega}{c} \tag{18.42b}$$

The first root yields a wave moving with the gravity wave speed to the west, and this yields as we have seen before an unbounded solution for the zonal velocity in y and must be rejected. The other root does yield a bounded solution. At low frequencies it looks like a Rossby wave; that is

$$\omega \approx -\frac{\beta}{k}$$

while for large frequencies, it looks like a pure gravity wave, $\omega \approx kc$.

Thus, there are two classes of solutions for each wave number. There is a set of higher frequency modes similar to the mid-latitude Poincaré modes and a set of low frequency modes corresponding to the Rossby modes. In addition, there is the Kelvin mode that exists only in its eastward traveling form. A single wave, which is discussed above, is often called the *mixed Rossby-gravity wave or the Yanai wave*, which straddles the two wave types. Each mode corresponding to a different j index goes along with the eigenfunction, $\psi_j(y/L_{eq})$ for its V-field except the Kelvin mode that has only a zonal velocity whose shape is given by ψ_0, the Gaussian.

The full dispersion relation is shown in Fig. 18.2. It is standard practice for the equatorial problem to consider only positive frequencies and to let the x-wave number run over positive and negative values.

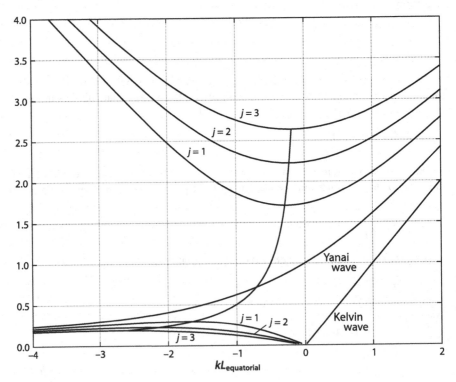

Fig. 18.2. The dispersion diagram for equatorial waves. The line bending upwards connects the extrema in the curve of frequency vs. wave number

In Fig. 18.2, the wave number is scaled with the equatorial deformation radius and the frequency is scaled with the characteristic time scale so that the frequency is given in units $\sqrt{\beta c}$. Note that both the Yanai wave and the Kelvin wave have only positive group velocity in the x-direction. Indeed from Eq. 18.37 it is easy to show that

$$\frac{\partial \omega}{\partial k}\left[\frac{2\omega}{\beta c}+\frac{kc}{\omega^2}\right]=c\left[\frac{2k}{\beta}+\frac{1}{\omega}\right] \tag{18.43}$$

or

$$\frac{\partial \omega}{\partial k}=\frac{c\left[\dfrac{2k\omega}{\beta}+1\right]}{\left[\dfrac{2\omega^2}{\beta c}+\dfrac{kc}{\omega}\right]} \tag{18.44}$$

The group velocity *vanishes* on the line

$$\omega=-\frac{\beta}{2k} \tag{18.45}$$

This line in the (ω,k)-plane separates westward from eastward group speeds and also marks the locus of the extrema in the frequency wave number plane.

If we insert the above condition in the dispersion relation, we obtain the value of the *minimum Poincaré frequency for each j*:

$$\omega_{\min}=\left(\frac{\beta c}{2}\right)^{1/2}\left[(j+1)^{1/2}+j^{1/2}\right], \quad \text{Poincarè} \tag{18.46a}$$

and

$$\omega_{\max}=\left(\frac{\beta c}{2}\right)^{1/2}\left[(j+1)^{1/2}-j^{1/2}\right], \quad \text{Rossby} \tag{18.46b}$$

For each j, these give the points of reversal of the sign of the group velocity. Note that the minimum of the Poincaré frequency is not at $k = 0$ but slightly displaced to negative k as a consequence of the beta effect. Note that the difference between the minimum Poincaré frequency for $j = 1$ and the maximum Rossby wave frequency for $j = 1$ satisfy

$$\frac{\omega_{\min\text{Poincarè}}}{\omega_{\max\text{Rossby}}}=\frac{\sqrt{2}+1}{\sqrt{2}-1}=5.828\ldots \tag{18.47}$$

so that both groups of waves are in the same range of parameter space. Any attempt to plot the Rossby waves and Poincaré waves of the same diagram in mid-latitude would be nearly impossible, since the frequencies are so disparate in size (this is, after all, the basis of quasi-geostrophy in mid-latitudes).

For each j, there is a corresponding eigenfunction for the V-field from which the zonal velocity field can be calculated, i.e.,

$$V_j = A_j \psi_j(\xi) e^{i(kx - \omega_j t)} \tag{18.48a}$$

$$U_j = A_j i(\beta c)^{1/2} \left[\left(\frac{j}{2} \right)^{1/2} \frac{\psi_{j-1}(\xi)}{[\omega + kc]} + \left(\frac{j+1}{2} \right)^{1/2} \frac{\psi_{j+1}(\xi)}{[\omega - kc]} \right] \tag{18.48b}$$

and this holds for the Poincaré, Rossby and Yanai waves. It is left to the student to work out the corresponding free surface elevations. The Kelvin wave of course has only a zonal component, and its amplitude is proportional to $\psi_0(\xi)$. The fact that each j mode consists of two eigenfunctions, ψ for j's one greater and one less than the j for V, renders the reflection problem rather complex. That, plus the physical fact that the Kelvin mode only exists in its eastward form makes the reflection problem from the eastern and western boundaries quite different, and the student is referred to the references given above for a detailed description of that problem.

Finally, we recall that each of the solutions above represents the contribution of a particular vertical mode with mode number n. Hence, each frequency and eigenfunction really should carry two indices, one for its horizontal structure (j) and one for its vertical structure (n).

Stratified Quasi-Geostrophic Motion and Instability Waves

We return from our brief visit to the equator and investigate the low frequency motions in mid-latitudes that occur in a stratified fluid. The motion we consider will be in near (quasi-)geostrophic balance, but we will develop the equations in an informal, heuristic way, leaning heavily on the formal analysis of Lecture 15. We will also employ the beta-plane approximation so that we are assuming that two parameters, $\varepsilon = U/f_0 L$, $b = \beta L / f_0$, are both small. That being the case, the lowest order balances in the horizontal momentum equation imply that

$$u = -\frac{p_y}{\rho_0 f_0} \equiv -\psi_y \tag{19.1a}$$

$$v = \frac{p_x}{\rho_0 f_0} \equiv \psi_x \tag{19.1b}$$

Note the beta-plane use of the constant reference value of the Coriolis parameter. As a consequence of that balance, it follows that at $O(1)$, the horizontal velocity is non-divergent, so that for an incompressible fluid,

$$\frac{\partial w}{\partial z} = O(\varepsilon, b) \tag{19.2}$$

If w vanishes at any z at the lower or upper boundary or approximately vanishes there, it follows that w itself is small. Indeed, w is smaller by a factor of ε or b compared to its geometrical scaling UD/L. In that case, the vorticity equation that arises at order ε can be written

$$\frac{\partial \zeta}{\partial t} + u\frac{\partial \zeta}{\partial x} + v\frac{\partial \zeta}{\partial y} + \beta v = f_0 \frac{\partial w}{\partial z} \tag{19.3}$$

In the vorticity equation, the contribution to the advection of vorticity due to w is negligible, since w is of a higher order in Rossby number than u and v, but its influence is felt by the stretching term on the right-hand side. Small as w is, the weak stretching is amplified by the large Coriolis parameter, the planetary vorticity, which is $O(\varepsilon^{-1})$ larger than the relative vorticity, which makes up for the smallness of w in the vorticity budget. In the above,

$$\zeta = v_x - u_y = \nabla^2 \psi \tag{19.4}$$

At the same time, the motion, which is assumed as usual to be adiabatic, satisfies

$$\frac{\partial \rho}{\partial t} + u \frac{\partial \rho}{\partial x} + v \frac{\partial \rho}{\partial y} + w \frac{\partial \rho_0}{\partial z} = 0 \qquad (19.5)$$

Again, the vertical velocity is negligible in providing a contribution to the advection of the perturbation density, but it does enter the advection term by its contribution to the advection of the large background density gradient, $\partial \rho_0 / \partial z$ (large, that is, with respect to the density gradients associated with the motion; we shall still assume that the background density varies *slowly compared to the vertical scale of the motion*).

Using the hydrostatic equation and the standard definition of the buoyancy frequency allows us to write the density equation as

$$\frac{d}{dt}\left(\frac{\partial p}{\partial z}\right) + w N^2 \rho_0 = 0 \qquad (19.6a)$$

$$\frac{d}{dt} = \frac{\partial}{\partial t} + u \frac{\partial}{\partial x} + v \frac{\partial}{\partial y} \qquad (19.6b)$$

$$N^2 = -\frac{g}{\rho_0}\frac{d\rho_0}{dz} \qquad (19.6c)$$

Eliminating w between the adiabatic equation and the vorticity equation and taking care to use the properties of the geostrophic velocity, we obtain as the governing equation for the geostrophic stream function

$$\left[\frac{\partial}{\partial t} + \psi_x \frac{\partial}{\partial y} - \psi_y \frac{\partial}{\partial x}\right]\left[\nabla^2 \psi + \frac{\partial}{\partial z}\left(\frac{f_0^2}{N^2}\frac{\partial \psi}{\partial z}\right)\right] + \beta \psi_x = 0 \qquad (19.7)$$

This is the *quasi-geostrophic potential vorticity equation (qgpve)*. It is important to note that although the motion is three-dimensional, i.e., a function of x, y and z, the advective term in the equation only reflects the effects of horizontal advection due to the rotation-induced smallness of w. A more systematic derivation is given in Pedlosky (1987).

At the lower boundary, the kinematic condition is

$$w = u h_{b_x} + v h_{b_y} = -\frac{1}{N^2 \rho_0}\frac{d}{dt}\frac{\partial p}{\partial z} \qquad (19.8)$$

Using the geostrophic relations for the horizontal velocities and the relation between p and ψ, we obtain for the boundary condition at $z = -D$

$$\frac{\partial}{\partial t}\frac{\partial \psi}{\partial z} + J(\psi, \psi_z) + \frac{N^2}{f_0} J(\psi, h_b) = 0 \qquad (19.9a)$$

$$J(a,b) \equiv a_x b_y - a_y b_x \qquad (19.9b)$$

Ignoring the deviation of the free surface with respect to that of the internal iso-pycnals (the same discussion as for internal waves), the boundary condition at $z = 0$ is just $w = 0$, which in terms of the geostrophic stream function is

$$\frac{\partial}{\partial t}\frac{\partial \psi}{\partial z} + J(\psi, \psi_z) = 0, \quad z = 0 \tag{19.20}$$

Let's examine that approximation a bit more carefully. From the adiabatic equation, the characteristic size of w generated within the fluid is of the order

$$w_{int} = O\left(\frac{UfUL}{LN^2D}\right) \tag{19.20}$$

where we have used the geostrophic scaling for p and the scaling U/L for the advective time derivative. On the other hand, the vertical velocity at the free upper surface will be of the order

$$w_{z=0} = d\eta/dt = O\left(\frac{Up}{L\rho_0 g}\right) = O\left(\frac{UfUL}{L\rho_0 g}\right) \tag{19.21}$$

The ratio is of the order

$$\frac{w_{z=0}}{w_{int}} = \frac{N^2 D}{g} \ll 1 \tag{19.22}$$

and so to the lowest order, w is zero at the free surface, which is the condition used above.

Let's look first for *baroclinic Rossby waves*. Let the bottom be flat and assume the motion is small amplitude so that we can linearize the dynamics. The problem then becomes

$$\frac{\partial}{\partial t}\left[\nabla^2 \psi + \frac{\partial}{\partial z}\left(\frac{f_0^2}{N^2}\frac{\partial \psi}{\partial z}\right)\right] + \beta\frac{\partial \psi}{\partial x} = 0 \tag{19.23a}$$

$$\frac{\partial^2 \psi}{\partial z \partial t} = 0 \tag{19.23b}$$

$$z = 0, -D \tag{19.23c}$$

We can find plane wave solutions in the form

$$\psi = A e^{i(kx + ly - \omega t)}\Phi(z) \tag{19.24}$$

where Φ satisfies the ordinary differential equation

$$\frac{d}{dz}\left(\frac{f_0^2}{N^2}\frac{d\Phi}{dz}\right) + \left\{\frac{\beta k}{(-\omega)} - k^2 - l^2\right\}\Phi = 0 \tag{19.25a}$$

$$\frac{d\Phi}{dz} = 0 \tag{19.25b}$$

$$z = 0, -D \tag{19.25c}$$

This is a standard Sturm-Liouville eigenvalue problem, and indeed it is the same problem we discussed for the vertical structure equation for Laplace's tidal equations. For the simple case where N is a constant, the solutions can be found immediately as

$$\Phi(z) = \cos(n\pi z / D), \quad n = 0,1,2,\ldots \tag{19.26}$$

(note that $n = 0$ is a nontrivial case) from which it follows that for each n,

$$\omega_n = -\frac{\beta k}{k^2 + l^2 + \dfrac{n^2 \pi^2}{L_D^2}}, \quad L_D = \frac{ND}{f_0} \tag{19.27}$$

The $n = 0$ mode is the barotropic mode. The horizontal motion is independent of z. The higher modes, each with a lower frequency for the same horizontal wave number, have zero vertically averaged horizontal velocity. For each n, the dispersion relation is exactly what we found for a homogeneous layer of fluid, except that now the term involving the deformation radius is the baroclinic deformation radius, and the barotropic mode is approximated by the limit where the deformation radius of the *free surface* is considered infinitely large compared to the L_D given above. This is the same approximation that allowed us to ignore w at $z = 0$. Again, mode by vertical mode, we can apply all the results of our investigations of the Rossby wave in a homogeneous layer to each vertical baroclinic mode.

Topographic Waves in a Stratified Fluid

Consider now the case where N is again constant, but a bottom slope exists and we ignore the beta effect. This last condition implies that the horizontal length scale is small enough that in the linear vorticity balance, $\omega K^2 \gg \beta k$, which we must check after the fact. In this case, the potential vorticity equation (linearized) is merely, for the same periodic plane wave in x and y,

$$-i\omega \left[\frac{d^2 \Phi}{dz^2} - \frac{N^2}{f_0^2} K^2 \Phi \right] = 0, \quad K^2 = k^2 + l^2 \tag{19.28}$$

We assume for simplicity that the bottom is sloping uniformly in the y-direction and that the upper surface at $z = 0$ is very far away (we have to quantify this idea shortly) so that the region can be considered infinite in z. Then the lower boundary condition is

$$-i\omega \Phi_z + ik \frac{N^2}{f_0} h_{b_y} \Phi = 0, \quad z = -D \tag{19.29}$$

The solution that decays away from the lower boundary and so remains finite with distance from the lower boundary is

$$\Phi = A e^{-KN(z+D)/f_0} \tag{19.30}$$

(the student should now show that the condition that the upper surface appears to be infinitely far away from the lower boundary is simply $KL_D \gg 1$, i.e., that the wavelength be short compared to the deformation radius).

Using this solution in the lower boundary condition yields the dispersion relation

$$\omega = -\frac{kh_{b_y}N}{K} \tag{19.31}$$

This is actually a very remarkable result. It has some similarities to the dispersion relation for the Kelvin wave. Here we have a single boundary at $z = -D$ and a wave with a single direction of propagation. The frequency, as in the case of the Kelvin wave, is *independent of the rotation*. On the other hand, again like the Kelvin wave, the trapping scale depends on f; only now, the trapping increases as f decreases. This bottom trapped wave has a vertical trapping scale δ such that if λ is the wavelength, $f\lambda / N\delta = 1$. Another way to look at the wave is to note that the bottom slope introduces a *topographic* beta effect:

$$\beta_{topog} = \frac{f_0 h_{b_y}}{D}$$

in terms of which

$$\omega = -\beta_{topog}\frac{kL_D}{K}$$

which has something of the character of a Rossby wave.

Waves in the Presence of a Mean Flow

Instead of linearizing about a state of rest, let's return to the full, quasi-geostrophic potential vorticity equation

$$\left[\frac{\partial}{\partial t} + \psi_x\frac{\partial}{\partial y} - \psi_y\frac{\partial}{\partial x}\right]\left[\nabla^2\psi + \frac{\partial}{\partial z}\left(\frac{f_0^2}{N^2}\frac{\partial\psi}{\partial z}\right)\right] + \beta\psi_x = 0 \tag{19.32}$$

and imagine that the wave is embedded in a mean zonal flow. That is, we will write the stream function as $\Psi(y, z)$, which represents a mean zonal flow *that is an exact solution of the qgpve*, and add to it a wave perturbation so that

$$\psi = \Psi(y, z) + \varphi(x, y, z, t) \tag{19.33}$$

Note that in the basic wave-free state, the zonal flow and the accompanying density anomaly are

$$U(y, z) = -\Psi_y \tag{19.34a}$$

$$\bar{\rho}(y, z) = -\frac{f_0\rho_0}{g}\Psi_z \tag{19.34b}$$

From these definitions or equivalently the thermal wind equation,

$$U_z = -\frac{g}{f_0 \rho_0} \frac{\partial \overline{\rho}}{\partial y} \tag{19.34c}$$

$$\left[\frac{\partial}{\partial t} + U\frac{\partial}{\partial x}\right]\left[\nabla^2\varphi + \frac{\partial}{\partial z}\left(\frac{f_0^2}{N^2}\frac{\partial\varphi}{\partial z}\right)\right] + \frac{\partial\varphi}{\partial x}\frac{\partial}{\partial y}\left[\Psi_{yy} + \frac{\partial}{\partial z}\left(\frac{f_0^2}{N^2}\frac{\partial\Psi}{\partial z}\right)\right] + \beta\varphi_x = 0 \tag{19.35a}$$

or

$$\left[\frac{\partial}{\partial t} + U\frac{\partial}{\partial x}\right]\left[\nabla^2\varphi + \frac{\partial}{\partial z}\left(\frac{f_0^2}{N^2}\frac{\partial\varphi}{\partial z}\right)\right] + \frac{\partial\varphi}{\partial x}\left[\beta - U_{yy} - \frac{\partial}{\partial z}\left(\frac{f_0^2}{N^2}\frac{\partial U}{\partial z}\right)\right] = 0 \tag{19.35b}$$

This equation is the perturbation form of the qgpve. The presence of the mean flow has produced two very important changes. First, the local time derivative has been changed to a linearized form of the advective derivative in which the additional term

$$U\frac{\partial}{\partial x}$$

represents the advection by the mean flow. Equally important (if not more so) is the fact that the planetary gradient of vorticity, β, is now supplemented by the contribution of the mean flow to the potential vorticity gradient of the basic state. That is, the meridional potential vorticity gradient is now

$$\frac{\partial\overline{q}}{\partial y} = \beta - U_{yy} - \frac{\partial}{\partial z}\left(\frac{f_0^2}{N^2}U_z\right) \tag{19.36}$$

This is analogous to the way, for a homogeneous fluid, the bottom topography supplements the beta effect to provide an altered potential vorticity gradient in which the wave propagates. However, as we shall see, the effect of the mean flow can do more than simply alter the frequency.

The boundary condition at $z = -D$ in this linearized problem becomes

$$\left(\frac{\partial}{\partial t} + U\frac{\partial}{\partial x}\right)\varphi_z + \varphi_x\Psi_{zy} + \frac{N^2}{f_0}\varphi_x h_{b_y} = 0, \quad z = -D \quad \text{or} \tag{19.37a}$$

$$\left(\frac{\partial}{\partial t} + U\frac{\partial}{\partial x}\right)\varphi_z + \varphi_x\left[-U_z + \frac{N^2}{f_0}h_{b_y}\right] = 0, \quad z = -D \tag{19.37b}$$

The last term in the square brackets can be written in a rather suggestive form:

$$\frac{N^2}{f_0}\left[h_{b_y} - U_z f_0 / N^2\right] = \frac{N^2}{f_0}\left[h_{b_y} - \frac{g}{f_0}\frac{\overline{\rho}_y}{(-g\rho_{0_z})}f_0\right]$$

$$= \frac{N^2}{f_0}\left[h_{b_y} + \frac{\overline{\rho}_y}{\rho_{0_z}}\right] = \frac{N^2}{f_0}\left[h_{b_y} - \left(\frac{\partial z}{\partial y}\right)_\rho\right] \tag{19.38}$$

so that the boundary condition contains terms involving the *difference* between the slope of the boundary and the slope of the basic state's isopycnal surfaces as they intersect the boundary. At the upper boundary, which we assume is flat,

$$\left(\frac{\partial}{\partial t}+U\frac{\partial}{\partial x}\right)\varphi_z+\varphi_x[-U_z]=0, \quad z=0 \tag{19.39}$$

Boundary Waves in a Stratified Fluid

Consider the situation in which for simplicity we ignore beta completely and suppose that the bottom boundary is flat, i.e., $h_b = 0$. Instead we will consider the case where the mean flow in the x-direction is *sheared in the vertical* so that

$$U=U_0+U_z\{z+D\} \tag{19.40}$$

where U_z is a constant, i.e., a flow with constant vertical shear. This is supported by a horizontal density gradient and hence a sloping density surface in the y-z-plane. Instead of the bottom sloping and the basic state density surfaces being flat, as in the case of the bottom trapped topographic wave we studied earlier, we now have the bottom flat and the density surfaces sloping. From the form of the boundary condition, however, these might have some equivalence. Let's see. For the case where N is constant and where the lower and upper boundaries are well-separated (in the sense of the topographic boundary wave discussed above), the qgpve is again, for Φ,

$$(-i\omega+Uik)\left[\frac{d^2\Phi}{dz^2}-\frac{N^2}{f_0^2}K^2\Phi\right]=0, \quad K^2=k^2+l^2 \tag{19.41}$$

leading again to the interior solution:

$$\varphi = Ae^{i(kx+ly-\omega t)}e^{-KN(z+D)/f_0} \tag{19.42}$$

The boundary condition at $z = -D$ now yields the relation

$$-(U_0ik-i\omega)KN/f_0-ikU_z=0 \quad \text{or} \tag{19.43a}$$

$$\omega=U_0k+\frac{k}{K}\frac{f_0}{N}U_z \tag{19.43b}$$

$$c=\frac{\omega}{k}=U_0+\frac{f_0U_z}{KN} \tag{19.43c}$$

The propagation consists of two parts. The first is a simple advection by U_0, which is the basic state velocity at $z = -D$. The more interesting contribution is from the vertical shear, or equivalently, the slope of the isopycnals relative to the lower surface. Indeed, the result for the frequency can be written as

$$\omega-U_0k=\frac{k(\partial z/\partial y)_\rho}{K}N \tag{19.44}$$

which should be compared to the relation for the bottom trapped topographic wave. In this simple case, this shows the equivalence between the sloping isopycnals and the sloping surface (the change in sign is prefigured by the differing signs in the boundary term)

$$\frac{N^2}{f_0}\left[h_{b_y}-\left(\frac{\partial z}{\partial y}\right)_\rho\right] \tag{19.45}$$

Now let's instead consider a wave localized near the *upper* boundary. The potential vorticity equation is the same, but the solution decaying *away* from the boundary is

$$\varphi = A e^{i(kx+ly-\omega t)}e^{zKN/f_0} \tag{19.46}$$

The boundary condition on $z = 0$ now yields

$$\left[(U_0+U_zD)ik-i\omega\right]\frac{KN}{f_0}-U_zik=0 \tag{19.47a}$$

$$\Rightarrow \omega=(U_0+U_zD)k-\frac{kU_z}{K}\frac{f_0}{N} \tag{19.47b}$$

Comparing this result to the case where the wave is trapped near the lower boundary, we see two differences. First, the advective velocity is different because the shear makes the advecting velocity larger at $z = 0$ (assuming positive shear). Second, the intrinsic frequency, i.e., the frequency seen by an observer moving with the local basic flow, had *the opposite sign* compared to the former case. It is the slope of the isopycnals relative to the boundary, and this has changed from the previous situation.

It is interesting to ask whether there is any wave number for which the two frequencies or the two phase speeds of these apparently independent waves could be equal. If that were the case, it could be possible for the two waves to effectively interact. Equating the two phase speeds in the two cases leads to

$$(U_0+U_zD)-\frac{U_z}{K}\frac{f_0}{N}=U_0+\frac{U_z}{K}\frac{f_0}{N} \tag{19.48}$$

$$\Rightarrow K=2\frac{f_0}{ND}=\frac{2}{L_D}$$

When the wave number is twice the inverse of the deformation radius, both boundary waves, one moving towards positive x *relative to the local flow at its boundary* and one moving towards negative x *relative to the local flow at its boundary*, are moving at the same speed relative to a fixed frame. In that case, we might wonder whether a particular mode can be produced from the interaction of these two waves. Note that when $KL_D = O(1)$, the assumption that the two boundaries are well-separated fails, and we have to consider the solution from first principles. A surprise results when we do.

Baroclinic Instability and the Eady Model

We return to our stratified layer with the shear flow previously described. The layer has an overall thickness D, and N is constant as well as the shear U_z. Again, we ignore the beta effect. The algebra is a bit more standard if we move the position of the origin in z to the lower boundary so that $0 \le z \le D$. The basic flow is thus

$$U = U_0 + U_z z \tag{19.49}$$

The boundary conditions are

$$\left(-i\omega + ik(U_0 + U_z z)\right)\Phi_z - U_z ik\Phi = 0 \tag{19.50a}$$

$$z = 0, D \tag{19.50b}$$

while the equation for Φ is

$$\frac{d^2\Phi}{dz^2} - \frac{N^2}{f_0^2}K^2\Phi = 0 \tag{19.51}$$

The general solution for Φ is

$$\Phi = A\cosh(\mu z) + B\sinh(\mu z) \tag{19.52a}$$

$$\mu \equiv \frac{NK}{f_0} \tag{19.52b}$$

Applying the boundary conditions at $z = 0$ and $z = D$ yields two equations for A and B:

$$z = 0, \quad -\tilde{c}\mu B - \mu A = 0 \quad \text{and} \tag{19.53a}$$

$$z = D, \quad (U_z D - \tilde{c})\left[\mu A \sinh(\mu D) + \mu B \cosh(\mu D)\right] \\ -U_z\left[A\cosh(\mu D) + B\sinh(\mu D)\right] = 0 \tag{19.53b}$$

where

$$\tilde{c} = \frac{\omega}{k} - U_0 \tag{19.53c}$$

The equations above are two homogeneous, linear, algebraic equations for the constants A and B. The condition for nontrivial solutions is that the determinant of the coefficients vanishes. This yields a quadratic equation for $\tilde{c}$:

$$\tilde{c}^2 - U_z D\tilde{c} + U_z^2\left(\frac{D}{\mu}\coth(\mu D) - \frac{1}{\mu^2}\right) = 0 \tag{19.54}$$

This yields two solutions:

$$\tilde{c} = \frac{U_z D}{2} \pm \frac{U_z D}{2}\left[1 - 4\frac{\coth(\mu D)}{(\mu D)} + 4\frac{1}{(\mu D)^2}\right]^{1/2} \tag{19.55a}$$

or

$$\tilde{c} = \frac{U_z D}{2} \pm \frac{U_z}{\mu}\left[\frac{(\mu D)^2}{4} - \mu D\coth(\mu D) + 1\right]^{1/2} \tag{19.55b}$$

The useful identity

$$\coth x = \frac{1}{2}\{\tanh(x/2) + \coth(x/2)\}$$

finally allows us to write the equation for the phase speed as

$$c = \frac{\omega}{k} = U_0 + \frac{U_z D}{2} \pm \frac{U_z}{\mu}\left[\left(\frac{\mu D}{2} - \coth(\frac{\mu D}{2})\right)\left(\frac{\mu D}{2} - \tanh(\frac{\mu D}{2})\right)\right]^{1/2} \tag{19.56}$$

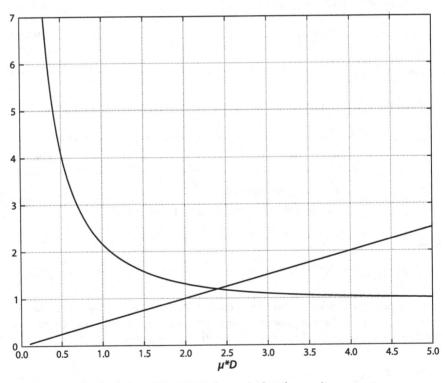

Fig. 19.1. A graphical solution of Eq. 19.57. Each term is plotted versus its argument

Since $x \geq \tanh x$ for all x, the second factor in the square bracket is always positive. The first factor will change sign where

$$\mu D/2 = \coth\left(\frac{\mu D}{2}\right) \qquad (19.57)$$

The graphical construction of the intersection (Fig. 19.1) shows that the critical value of $\mu D \approx 2.3994$.

Note that this corresponds to a value of $K = 2.3997 / L_D$ rather close to the heuristically motivated value from the previous discussion. For wave numbers less than this critical K, the frequency will be *complex*.

When c is complex, i.e., when $c = c_r + ic_i$, the behavior in time consists of an oscillation and an exponential growth for positive c_i, i.e., the time factor becomes $e^{-ikct} = e^{-ikc_r t} e^{kc_i t}$ with a *growth rate*

$$\omega_i = \frac{kU_z}{\mu}\left[(\coth(\mu D/2) - \mu D/2)(\mu D/2 - \tanh(\mu D/2))\right]^{1/2} \qquad (19.58)$$

Figure 19.2 below shows the real part of the phase speed measured with respect to U_0 and scaled by $U_z D$. The dotted line shows the imaginary part of c also scaled with $U_z D$,

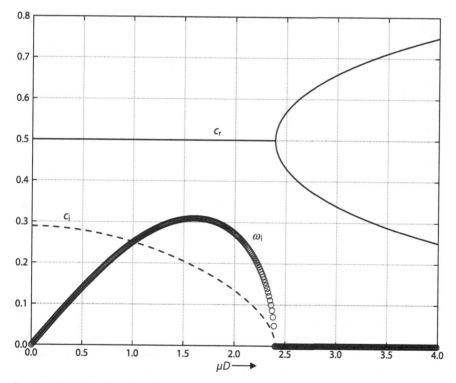

Fig. 19.2. The Eady dispersion relation K scaled on deformation radius. The *thin curves* show the real and imaginary parts of the phase speed. The *heavy line* made of "o's" shows the growth rate

while the imaginary part of the frequency, the *growth rate* scaled by U_z or the shear, is shown to be the line formed by the circles. The figure is drawn for the case where $l = 0$.

Note that for each wave number for which a positive imaginary part of c exists, there is another solution with the same real part of c but whose imaginary part is negative. This follows from the fact that the equation of the perturbation field is real, so that if Φ is a solution with eigenvalue c, its complex conjugate Φ^* will be a solution with eigenvalue c^*. Since the dynamics is inviscid and thus reversible in time, the physics must include the possibility for a cunningly chosen initial condition to return a disturbance to zero amplitude (exponentially slowly).

The model described above was initially described by E. T. Eady (1949). This explanation and the paper by Jule Charney (1947) were the first to correctly describe the instability process now known as *baroclinic instability* of which the Eady model is perhaps the simplest example. The accomplishments of both these independent analyses are staggering. Not only did Eady and Charney, who were working independently, correctly identify the physical process responsible for synoptic scale waves in the atmosphere (and ocean), but they had to derive a version of quasi-geostrophy at the same time. For those of you starting graduate school, it will give you a standard to strive for to know that this represented Charney's Ph.D. thesis.

For K greater than the critical value $K_c = 2.3999 / L_D$, both roots for the phase speed are real. As K gets very large, each root approaches the value of the zonal velocity at

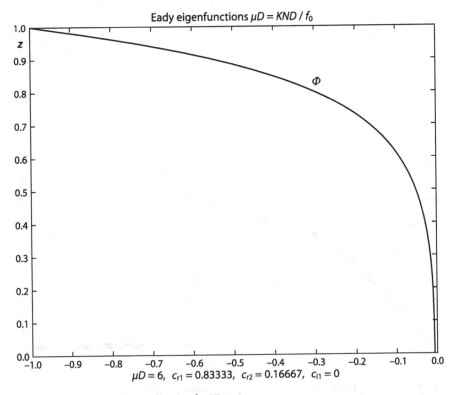

Fig. 19.3. The stable Eady eigenfunction for $KL_D = 6$

one of the boundaries at $z = 0$ or $z = 1$, and the eigenfunction resembles the trapped boundary wave of the semi-infinite interval model. Figure 19.3 shows an example when $KL_D = 6$. The eigenfunction with $\tilde{c} / (U_z D)$ about 0.8 is shown. Note its intensification near the upper boundary.

In this limit, there is no interaction between the upper and lower boundaries and the wave is stable. We shall see later a theorem that will explain why, in the Eady model, such an interaction is necessary for instability.

As K approaches its critical value from above, the two roots for c coalesce. For K less than the critical value, there are two roots which are complex conjugates. The growth rate is

$$\omega_i = kc_i(K) = kc_i\left(\sqrt{k^2 + l^2} \right) \tag{19.59}$$

That is, the complex phase speed, as we can see from the original eigenvalue problem is a function *only* of the total wave number. The growth rate is the imaginary part of that phase speed multiplied by the x-wave number, i.e., by the component of the wave vector in the direction of the basic velocity. The largest growth rate will therefore occur for a given K when k is largest, i.e., when the y-wave number, l, is zero. In the figure showing the growth rate, I have chosen the case where $l = 0$. The maximum growth rate occurs for k on the order of $1.6 / L_D$, which gives a quarter wavelength of just under L_D itself. This is the basic explanation for the presence in both the atmosphere and the ocean of synoptic scale eddies with the scale, preferentially, of the deformation radius.

The fact that the instability is maximized for $l = 0$ is related to the energy source for the waves. Since the motion is horizontally divergent to the lowest order (geostrophic), the perturbation velocity is perpendicular to the wave vector. With the wave vector oriented in the x-direction, the perturbation velocity will be directed *across* the current in the y-direction.

Now from the *thermal wind relation,*

$$\frac{\bar{\rho}_y}{\rho_0} = \frac{f_0 U_z}{g}$$

and using the relation

$$\frac{\rho_{0_z}}{\rho_0} = -\frac{N^2}{g}$$

it follows that the slope of the isopycnals in the basic state is

$$\left. \frac{\partial z}{\partial y} \right)_\rho = \frac{f_0 U_z}{N^2} \tag{19.60}$$

Motion in the y-direction will therefore move fluid down the density gradient, and the fluid motion has a chance to release the potential energy that is stored in the sloping density surfaces, a slope required to balance the vertical shear of the current in which the wave is embedded.

The important point to keep in mind is that these waves are *self-excited*; since they are unstable, the slightest perturbation of the basic flow will produce a spectrum of growing waves, and we anticipate that at least until the amplitude becomes large enough for nonlinearity to be important, we will see the *most* unstable wave dominate the spectrum. That is, we don't need an external forcing mechanism to produce the wave, in distinction to all the wave types we have discussed before. We will have to discuss more completely the energy source for the waves that is in the basic current, but it should be intuitively clear that the sloping density surfaces are a potential source of energy if the perturbations on average can level those surfaces releasing potential energy to perturbation kinetic energy.

For the unstable wave, the function Φ will be complex, since c is complex and A and B will thus be complex. It is useful to recognize this and write the stream function:

$$\psi = \Phi(z)e^{i(kx-\omega t)}$$
$$= |\Phi(z)| e^{i(kx+\alpha(z)-\omega t)} \tag{19.61}$$

In the above, we have separated the amplitude function Φ into its modulus and its phase

$$\alpha(z) = \arg(\Phi) \tag{19.62}$$

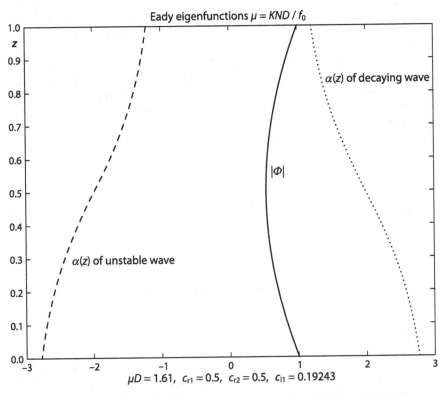

Fig. 19.4. The modulus (*solid*) and phase (*dashed*) for the stable ($c_i < 0$) and unstable ($c_i > 0$) waves. The phase of the unstable mode increases with height

A line of constant phase in the x-z-plane will be a line on which

$$ph = kx + \alpha(z) = \text{constant} \tag{19.63}$$

The slope of a line of constant phase is therefore

$$\left(\frac{\partial z}{\partial x}\right)_{ph} = -\frac{k}{d\alpha/dz} \tag{19.64}$$

Figure 19.4 shows the modulus of the eigenfunction and its phase for a wave number very near the wave number of maximum growth rate. Both the unstable wave and its complex conjugate are shown. Of course, the modulus of Φ is the same for both. Note that the function α increases with z for the unstable wave and decreases with z for the stable wave.

For the unstable wave, the fact that α increases with z means that *a line of constant phase of the unstable wave has a negative slope in the x-z-plane; that is, it leans against the current shear*. Intuitively, for a passive tracer we would expect isolines of the tracer to be pitched over in the direction of the shear. The unstable wave has an active structure, and to extract energy from the basic flow it must lean against the shear in the current.

Energy Equation
and Necessary Conditions for Instability

To get a better feeling for where the source of the instability is, it is useful to develop an equation for the perturbation energy for waves in the presence of a mean flow that contains both horizontal and vertical shear. This entire subject is enormous, and we will only scratch the surface in our discussion. The text by Gill (1982) and Pedlosky (1987) contain ample discussion for further reading.

We start with the governing equation for the linear perturbations derived in the last lecture:

$$\left[\frac{\partial}{\partial t}+U\frac{\partial}{\partial x}\right]\left[\nabla^2\varphi+\frac{\partial}{\partial z}\left(\frac{f_0^2}{N^2}\frac{\partial\varphi}{\partial z}\right)\right]+\frac{\partial\varphi}{\partial x}\left[\frac{\partial\overline{q}}{\partial y}\right]=0 \tag{20.1}$$

where the term in the last bracket is the potential vorticity gradient in the y-direction associated with the basic flow. It contains contributions from β, the relative vorticity gradient and the gradient of the thickness between isopycnal surface in the basic state

$$\frac{\partial\overline{q}}{\partial y}=\beta-U_{yy}-\frac{\partial}{\partial z}\left(\frac{f_0^2}{N^2}\frac{\partial U}{\partial z}\right)$$

$$=\beta-U_{yy}-f_0\frac{\partial}{\partial z}\left(\left\{\frac{\partial\overline{z}}{\partial y}\right\}_\rho\right) \tag{20.2}$$

where $\overline{z}(y)$ is the position of an isopycnal in the basic state.

To derive the energy equation, we follow the steps we took earlier in finding the energy flux vector for quasi-geostrophic Rossby waves. We multiply the qgpve by the stream function and manipulate the derivatives to work the form into a budget for the energy. The details are a good deal more tedious here, because U is a function of y and z and there are many "extra" terms. It is these terms that are the most illuminating. The details of the derivation will be left for the student. The result with no further approximation can be written

$$\frac{\partial}{\partial t}E+\nabla\cdot\vec{S}+\frac{\partial}{\partial z}\left(\varphi\frac{f_0^2}{N^2}\{\varphi_{zt}+U\varphi_{zx}\}\right)=-\varphi_x\varphi_y U_y-\varphi_x\varphi_z\frac{f_0^2}{N^2}U_z \tag{20.3a}$$

$$E=\frac{(\nabla\varphi)^2}{2}+\frac{f_0^2\varphi_z^2}{2N^2} \tag{20.3b}$$

$$\vec{S} = -\varphi \left[\left(\frac{\partial}{\partial t} + U \frac{\partial}{\partial x} \right) \nabla \varphi \right] + \hat{x} \left\{ -\frac{\partial \bar{q}}{\partial y} \frac{\varphi^2}{2} + UE + \varphi (\nabla \varphi \cdot \nabla U) \right\} + \varphi U \frac{\partial \nabla \varphi}{\partial x} \qquad (20.3c)$$

where $\hat{x}$ is a unit vector in the x-direction. The energy E is the sum of the kinetic energy and the potential energy. The second term in E is the potential energy, since

$$\frac{f_0^2 \varphi_z^2}{2N^2} = \frac{g^2 (\rho / \rho_0)^2}{2N^2} \qquad (20.4)$$

which we recognize from our discussion of internal waves as the representation of the potential energy in the wave field. The horizontal flux vector is similar to that which we found for Rossby waves. The local time derivative in the first term is replaced by the linearized advective derivative, and the beta term is replaced by the full potential vorticity gradient. This is supplemented by the advection of energy in the x-direction by the mean flow plus two other terms. These terms are more difficult to interpret easily, but they are related to corrections to the higher order work terms done by the geostrophic pressure correction. The horizontal divergence of this flux vector has its companion in the z-direction.

If the fluid is contained within solid walls in z and y so that the boundary conditions at $z = 0$ and $z = -D$ are as described in the previous lecture, and if the perturbation is either periodic in x or vanishes for large positive and negative x, then the volume integral of the flux terms will contribute no net term to the energy balance for the perturbation energy. This is really just a consequence of the definition of energy flux. The flux vector moves the energy from one place to another without creating or destroying energy.

However, there are two terms: these terms are on the right-hand side of the energy equation that *in general*, do not integrate to zero when the volume integral is carried out. The first of these is already familiar from our discussion of the energy flux in internal gravity waves in a mean current. Using a bracket to denote a volume integral, this term is

$$\langle \varphi_x \varphi_y U_y \rangle = -\langle u v U_y \rangle$$

and thus is the integral of the horizontal Reynolds stress times the horizontal shear of the basic current. If the perturbation carries larger values of zonal momentum to regions of lower momentum tending to smooth out the mean lateral shear, i.e., if when $U_y > 0$ and $v < 0$ we also have $u > 0$ so that the perturbations "remember" that they have come from a region of large zonal momentum compared to their destination, the mean shear will be flattened with a consequent increase in wave energy as the energy of the basic current is reduced. Such an energy transfer requiring only horizontal motions occurs in ordinary shear flow instability of a homogeneous fluid with lateral shear and is termed *barotropic instability*.

In the Eady model we discussed in the last lecture, the basic current has no horizontal shear so that this energy transformation process is absent. The remaining term on the right-hand side is the pertinent one for that process. Using the relation between geostrophic stream function and density perturbation,

$$\frac{f_0^2}{N^2}\varphi_x\varphi_z U_z = -v(\rho g)\frac{g\overline{\rho}_y}{\rho_0 N^2} = -v\rho g\left(\frac{\partial z}{\partial y}\right)_{\overline{\rho}} \tag{20.5}$$

This baroclinic energy transformation term is proportional to the transport in the y-direction, i.e., in the direction of the density gradient of perturbation density. If, for example $\overline{\rho}_y > 0$, $\Leftrightarrow U_z > 0$, on average (when integrated over the volume of the fluid) parcels moving to positive y carry a negative density anomaly and fluid elements moving from positive y carry a positive density anomaly. The product $-v\rho > 0$ and so the energy of the wave field will increase. That is, if the wave field produces a flux of density from regions of high to low density of the basic state, tending to smooth out the basic horizontal density gradient, this will flatten the slope of the mean isopycnal surfaces and release energy for the perturbations.

It is really a form of convection. In ordinary convection in which fluid is heated from below, energy is released by having warm, light fluid rise and cold, heavy, dense fluid sink. Here the situation is a bit more complex, but fluid coming from the region of larger mean density moving to smaller mean density will tend to sink as it moves laterally and vice-versa for the fluid moving in the opposite direction.

Notice that to have this transformation term positive, we need

$$0 < U_z\varphi_x\varphi_z = U_z\varphi_z^2\frac{\varphi_x}{\varphi_z} = -U_z\varphi_z^2\left(\frac{\partial z}{\partial x}\right)_{\varphi} \tag{20.6}$$

so that the product of the basic vertical shear multiplied by the slope of the isolines of constant ϕ in the x-z-plane must be negative. That is, the phase lines of constant perturbation of the geostrophic stream function must lean against the shear, as we already noted from the Eady model. Here we see that it is a necessity to release the potential energy locked up in the sloping isopycnal surfaces of the basic state. Figure 20.1 shows a cross-section in the x-z-plane of a growing Eady mode. The *solid lines* are the geostrophic stream function, and the *dashed curves* yield contours of perturbation density. Note that the former lean against the shear, and the latter lean with the shear. This phasing assures that on average the density flux is *down* the mean density gradient.

In both the ocean and the atmosphere, horizontal density gradients are sources of baroclinic eddy energy, and the eddies springing from the self-excited baroclinic unstable waves typically have scales of the order of the appropriate deformation radius.

Eady himself introduced a very simple argument to make plausible the convective nature of the instability. He suggested considering the virtual displacement of a fluid element in the y-z-plane. The isopycnals are sloping with an angle γ such that

$$\tan\gamma = \left(\frac{\partial z}{\partial y}\right)_{\overline{\rho}} = \frac{f_0 U_z}{N^2}$$

so that the slope is due to the existence of the vertical shear (Fig. 20.2).

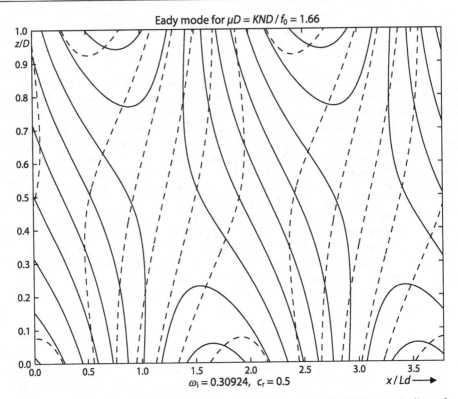

Fig. 20.1. A cross-section in the x-z-plane of the growing Eady mode. The *solid lines* are isolines of perturbation pressure, while the *dashed lines* show isolines of perturbation density

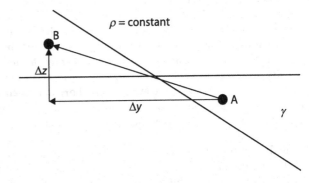

Fig. 20.2.
The Eady wedge of instability

Consider a displacement of a fluid parcel from point A to point B as indicated in the figure. Assuming the fluid parcel at A preserves its density when it arrives at B, it will arrive there with a density anomaly:

$$\delta\rho = \rho_A - \rho_B \tag{20.7}$$

But if the displacement is small,

$$\rho_B = \rho_A + \frac{\partial \rho}{\partial z}\Delta z + \frac{\partial \rho}{\partial y}\Delta y \qquad (20.8a)$$

∴

$$
\begin{aligned}
\delta \rho &= -\left(\frac{\partial \rho}{\partial z}\Delta z + \frac{\partial \rho}{\partial y}\Delta y \right) \\
&= -\frac{\partial \rho}{\partial z}\Delta z\left[1 + \frac{\dfrac{\partial \rho}{\partial y}\Big/\dfrac{\partial \rho}{\partial z}}{\dfrac{\Delta z}{\Delta y}} \right]
\end{aligned}
\qquad (20.8b)
$$

so that the anomaly of buoyancy will be

$$g\frac{\delta \rho}{\rho_0} = N^2 \Delta z\left[1 - \frac{z_y)_\rho}{\Delta z/\Delta y} \right] \qquad (20.9)$$

If Δy is zero or if the slope of the basic isopycnals is zero, this reduces to the result we obtained in reasoning out the restoring force for internal gravity waves. In that case, we had a restoring force (a positive density anomaly for a positive Δz) giving rise to a force proportional to Δz and with N^2 as the spring constant (per unit mass). Now, however, if

$$0 \le \frac{\Delta z}{\Delta y} \le \frac{\partial z}{\partial y}\bigg)_\rho$$

the buoyancy anomaly will be negative and the arriving fluid parcel will have lower density than its surroundings. The resulting buoyancy force will then encourage a continued displacement and the release of energy. That is, if the motion occurs so that *on average* the fluid elements slope upwards within a wedge determined by the slope of the density surfaces with respect to the horizontal, the gravitational energy available will power continued displacement rather than restoration to its initial position ⟶ instability. From this point of view, the instability is a type of slanted convection requiring vertical shear to yield the slope of the isopycnals and allowing the existence of *the wedge of instability.*

This simple explanation has been criticized (Heifetz et al. 1998), since the wave function is not a plane wave in the x-z-plane, so it is not possible to avoid considering the pressure perturbation in the force balance on the parcel. However, the basic argument on the basis of the buoyancy force is compelling and, I feel, illuminating. In Fig. 20.3 we show a snapshot of the v and w velocities in the y-z-plane at a particular value of x (quarter wavelength). The *solid lines* are the isopycnals, and the *arrows* show the trajectory in the y-z-plane.

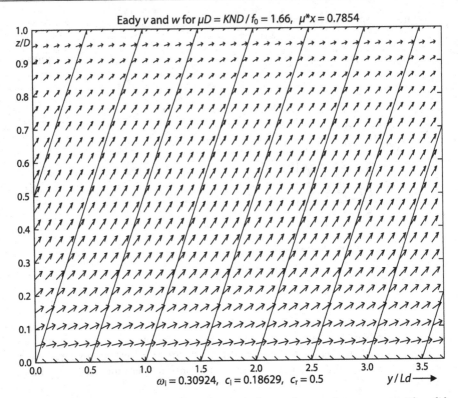

Fig. 20.3. The instantaneous perturbation velocities in the *y-z*-plane are shown as *arrows*. The *solid lines* are the basic state isopycnals

It is of interest to try, on the basis of our work up to this point and general dimensional analysis ideas, to estimate the characteristic growth rate of baroclinic, unstable waves. In the presence of a vertical shear, a layer of depth D might be expected to have the imaginary part of its phase speed to scale with $U_z D$. The growth rate would then be

$$\omega_i = k c_i = O(k D U_z) \tag{20.10}$$

and if the wavelength is of the order of the deformation radius,

$$k = O(f_0 / N D) \tag{20.11}$$

so that

$$\omega_i = O\left(\frac{f_0 U_z}{N}\right) = N\left(\frac{\partial z}{\partial y}\right)_\rho \tag{20.12}$$

(note in the last form its apparent independence of f).

For mid-ocean flows, we might estimate

$$U_z = \frac{1\,\text{cm s}^{-1}}{10^5\,\text{cm}}$$

$$N = 5\times10^{-3}\,\text{s}^{-1}$$

$$f_0 = 10^{-4}\,\text{s}^{-1}$$

$$\Rightarrow \omega_i = \frac{10^{-6}}{5}\,\text{s}^{-1}$$

which yields an e-folding time of about 60 days.

General Conditions for Instability

The Eady model is a very simple one and hardly realistic. The investigation of more realistic velocity structures usually requires considerable numerical work, and it is hard to make general statements. It is useful to have some a priori ideas of when self-excited waves can be expected within geostrophic flows. There are a series of theorems giving *necessary conditions for instability*. The student is referred to Chapter 7 of GFD for a detailed discussion. Here we present only the most well-known theorems. This class of theorem dates back to the original work of Lord Rayleigh.

Let us assume that our basic current is again directed in the zonal direction, but imagine that it is now a function of both y and z and that the beta effect is not negligible. It is important to note that once the current is not zonal, there are very few theorems that are directly applicable.

However, the case is of interest, and it may be provide some general picture of what is required even in the nonzonal case.

Again, if we look for plane waves in x (not y, since now the linearized potential vorticity equation has nonconstant coefficients in y) the governing equation

$$\left[\frac{\partial}{\partial t}+U\frac{\partial}{\partial x}\right]\left[\nabla^2\varphi+\frac{\partial}{\partial z}\left(\frac{f_0^2}{N^2}\frac{\partial\varphi}{\partial z}\right)\right]+\frac{\partial\varphi}{\partial x}\left[\frac{\partial\overline{q}}{\partial y}\right]=0 \tag{20.13}$$

admits solutions of the *normal mode* form

$$\varphi=\Psi(y,z)e^{ik(x-ct)}$$

where Ψ satisfies

$$(U-c)\left[\frac{\partial}{\partial z}\left(\frac{f_0^2}{N^2}\frac{\partial\Psi}{\partial z}\right)+\frac{\partial^2\Psi}{\partial y^2}-k^2\Psi\right]+\frac{\partial\overline{q}}{\partial y}\Psi=0 \tag{20.14}$$

subject to boundary conditions on the bottom (which we take to be $z=0$) and the top ($z=D$):

$$(U-c)\Psi_z + \Psi\left(-U_z + \frac{N^2}{f_0}h_{b_y}\right) = 0, \quad z = 0 \tag{20.15a}$$

$$(U-c)\Psi_z + \Psi(-U_z) = 0, \quad z = D \tag{20.15b}$$

and we insist that there exist two lateral boundaries at $y = \pm L$ where the disturbance vanishes. Of course, L could be infinite.

If we multiply the equation for Ψ by its complex conjugate after having divided by $(U-c)$, and if we then integrate over the region of the problem in the y-z-plane, we obtain, with the aid of the above boundary conditions

$$
\begin{aligned}
&-\int_0^D dz \int_{-L}^L dy \left[\frac{f_0^2}{N^2}|\Psi_z|^2 + |\Psi_y|^2 + k^2|\Psi|^2\right] \\
&+ \int_0^D dz \int_{-L}^L dy \left\{\frac{|\Psi|^2 \partial \bar{q}/\partial y}{U-c}\right\} + \int_{-L}^L dy \frac{f_0^2|\Psi|^2}{N^2(U-c)}\{U_z\}_{z=D} \\
&- \int_{-L}^L dy \frac{f_0^2|\Psi|^2}{N^2(U-c)}\left\{U_z - \frac{N^2}{f_0}h_{b_y}\right\}_{z=0} = 0
\end{aligned}
\tag{20.16}
$$

The first term in this integral condition is always real and negative definite. If c is complex, the remaining terms will have an imaginary part. Indeed, if we just write down the imaginary part of the above equation using

$$\frac{1}{(U-c)} = \frac{(U-c_r+ic_i)}{|U-c|^2} \tag{20.17}$$

we obtain

$$
c_i \left[
\begin{aligned}
&\int_0^D dz \int_{-L}^L dy \left\{\frac{|\Psi|^2 \partial \bar{q}/\partial y}{|U-c|^2}\right\} + \int_{-L}^L dy \frac{f_0^2|\Psi|^2}{N^2|U-c|^2}\{U_z\}_{z=D} \\
&- \int_{-L}^L dy \frac{f_0^2|\Psi|^2}{N^2|U-c|^2}\left\{U_z - \frac{N^2}{f_0}h_{b_y}\right\}_{z=0}
\end{aligned}
\right] = 0
\tag{20.18}
$$

For instability to occur, i.e., for the imaginary part of c to be different from zero, the collection of integrals in the square bracket must add to zero.

For example in the Eady problem, the potential vorticity gradient in the interior of the fluid is exactly zero, i.e.,

$$
\begin{aligned}
\frac{\partial \bar{q}}{\partial y} &= \beta - U_{yy} - \frac{\partial}{\partial z}\left(\frac{f_0^2}{N^2}\frac{\partial U}{\partial z}\right) \\
&= \beta - U_{yy} - f_0 \frac{\partial}{\partial z}\left(\left\{\frac{\partial \bar{z}}{\partial y}\right\}_\rho\right) \\
&= 0
\end{aligned}
\tag{20.19}
$$

For instability to occur, the two boundary terms must be able to cancel each other. For the Eady problem, there is no topography and U_z is positive at both boundaries so that the cancellation is possible. However, as we saw, it is necessary that the wave number be small enough so that the wave extends to both boundaries. If the eigenfunction were zero at one of the boundaries, only one of the boundary terms in the above constraint would survive, and it would be impossible to satisfy the condition for instability.

We noted earlier the term

$$U_z - \frac{N^2}{f_0} h_{b_y} \propto \left[\left(z_y \right)_\rho - h_{by} \right] \tag{20.20}$$

Thus, if the topography of the lower boundary were sloping upward more steeply than the isopycnals (which have constant slope in the Eady model), the contribution from the lower boundary term would add to that of the upper boundary, and instability would be impossible. Topography can thus eliminate the instability and stabilize the flow.

If both boundaries are boundaries of constant density, the boundary terms in the integral condition vanishes. In that case, for instability, the gradient of the potential vorticity must be both positive and negative in the y-z-plane. Potential vorticity of a single sign would be (in the absence of the boundary contributions) a stable distribution. The simple exemplar is, naturally, pure Rossby waves.

In Charney's model, there is no contribution from the upper boundary (it is infinitely far away) and the potential vorticity gradient is positive. Instability is possible because the positive contribution from the pv integral is cancelled by the contribution from the lower boundary. Note that these conditions are necessary conditions for instability, not sufficient conditions. It sometimes occurs that the necessary condition is met and the flow is still stable. There are very few useful sufficient conditions that can be found.

Wave–Mean Flow Interaction

We have been considering the dynamics of waves in this course and have remarked several times on the linearization restriction we have normally placed on the dynamics to make progress, and we have skirted rather completely the role of nonlinearity on the dynamics of the waves themselves. It is a difficult subject.

At the same time, even small amplitude waves, for which linear theory might be a good first approximation, can have an effect on the mean state of the medium through which the waves are propagating. If the waves have small amplitude, we would anticipate that since the fluxes of momentum and density by the waves are of $O(\text{amplitude})^2$, the effect on the mean will be similarly small. That doesn't mean that alteration is unimportant or uninteresting, and the calculation of that change often can give insight into how the waves can have an effect on the mean fields in which they are embedded. The role of waves in altering the mean is clearly of importance in questions concerning the general circulation or even current systems of smaller scales, e.g., coastal currents.

How then can we calculate the effect of waves on the mean field? This, too, is a very complicated subject, and in this lecture we will just touch on a special case but one which is both revealing and often used as a model for more general situations. We will consider the effects of waves on the mean for low-frequency large-scale motions governed by quasi-geostrophic dynamics. Even with these restrictions, the issue is complicated, and we will simplify further by considering mean states that correspond to zonal flows that are functions of y, z and t but that are independent of x. The waves, of course, will be functions of all three spatial variables. We will define the mean by the spatial average

$$\bar{P} = \lim_{X \to \infty} \int_{-X}^{X} P dx \qquad (21.1)$$

by an average in x where P is any dependent variable. It would appear that this definition is more suitable for atmospheric flows, and it is certainly true that the discussion that follows came first from the meteorological literature, but one can imagine strong currents that are nearly zonal such as the Gulf Stream after separation, the equatorial currents, etc. for which this is at least a sensible first approach.

The governing equation for the full system of waves and the mean flow is the potential vorticity equation, which in quasi-geostrophy is

$$\frac{\partial q}{\partial t}+u\frac{\partial q}{\partial x}+v\frac{\partial q}{\partial y}=0 \tag{21.2a}$$

$$u=-\psi_y \tag{21.2b}$$

$$v=\psi_x \tag{21.2c}$$

$$q=\nabla^2\psi+\frac{\partial}{\partial z}\left(\frac{f_0^2}{N^2}\frac{\partial\psi}{\partial z}\right)+\beta y \tag{21.2d}$$

We will assume that the wave field is periodic in x with no mean so that the x-average of any variable associated with the wave field will be zero. Furthermore, the x-average of the geostrophic meridional velocity must itself be zero, if the flow is periodic in x or independent of x.

We can therefore write all variables as a mean plus a wave part:

$$P=\overline{P}+P' \tag{21.3a}$$

$$\overline{P'}\equiv0 \tag{21.3b}$$

Noting first that the pv equation can be written

$$\frac{\partial q}{\partial t}+\frac{\partial uq}{\partial x}+\frac{\partial vq}{\partial y}=0 \tag{21.4}$$

since the geostrophic flow has zero horizontal divergence, an x-average of the pv equation yields, using $\overline{v}=0$,

$$\frac{\partial\overline{q}}{\partial t}=-\frac{\partial}{\partial y}(\overline{v'q'}) \tag{21.5}$$

The rate of change of the mean potential vorticity is given entirely in terms of the *potential vorticity flux of the waves*. Using the above equation, assuming the pv flux is known, one can simply calculate $\overline{q}$. Since the variable part of $\overline{q}$ satisfies the relation with the stream function

$$\overline{q}=\overline{\psi}_{yy}+\frac{\partial}{\partial z}\left(\frac{f_0^2}{N^2}\frac{\partial\overline{\psi}}{\partial z}\right) \tag{21.6}$$

it is with appropriate boundary conditions possible to invert to find $\overline{\psi}$, once $\overline{q}$ is known. In this sense, the problem of wave mean flow interaction is straightforward. First, one calculates the linear wave field. Then, one finds the average flux of pv in the wave field. Its meridional divergence yields the change in the mean pv. Once computed, that, with the inversion of the elliptic problem for stream function in terms of q, completes the specification of the change in the mean. Note that the forcing of that change is due *entirely* to the pv flux in the waves.

While the problem is formally complete, the approach leaves two important issues unclear. First, what can we say, a priori, about the pv flux in the wave field? Second, is there a more direct and physically intuitive way we can understand how the waves alter the mean momentum and density distribution other than the inversion of the above equation relating stream function and potential vorticity?

Let's take up the second question first. We note that

$$\overline{v'q'} = \overline{v'\left[\zeta' + \left(\frac{f_0^2}{N^2}\psi'_z\right)_z\right]}, \quad \zeta' = v'_x - u'_y \tag{21.7a}$$

$$\overline{v'\zeta'} = \overline{v'v'_x} - \overline{v'u'_y} = -\overline{v'u'_y} = -\overline{(v'u')}_y \tag{21.7b}$$

where we have repeatedly used the fact that $\overline{P_x} = 0$ for any variable P. With the geostrophic and hydrostatic approximations, this allows the pv wave flux to be written

$$\overline{v'q'} = -\overline{(v'u')}_y - \left(\overline{v'\rho'}\frac{gf_0}{\rho_0 N^2}\right)_z \tag{21.8}$$

The potential vorticity flux is therefore the divergence in the y-z-plane of the vector

$$\vec{F} = -\overline{u'v'}\vec{j} - \overline{v'\rho'}\frac{gf_0}{\rho_0 N^2}\vec{k} \tag{21.9}$$

where $\vec{j}$ and $\vec{k}$ are unit vectors in the y- and z-directions, respectively. The vector $\vec{F}$ is the *Eliassen and Palm (EP) flux vector*. Its horizontal component is the meridional wave flux of zonal momentum per unit mass, or equivalently the Reynolds stress, while its vertical component is, aside from a factor, the *horizontal meridional* density flux by the waves. The direction of $\vec{F}$ in the y-z-plane gives us an immediate sense of whether the meridional pv flux is due to momentum or density fluxes. For example, in the Eady problem, the resulting unstable waves would have a purely vertical EP vector.

Note that the mean pv equation is simply

$$\frac{\partial \overline{q}}{\partial t} = -\frac{\partial \nabla \cdot \vec{F}}{\partial y} \tag{21.10}$$

This is not apparently a great advance over our previous formulation, but it again emphasizes the point that the change in the mean fields will be due entirely (with the appropriate analysis of boundary effects skipped over here) to the divergence of the EP vector.

This somewhat new formulation is of great assistance when we examine the x-average of the momentum equation itself. If we take the x-average of the x-momentum equation and remember that the x-average of the *geostrophic* v is zero, we obtain

$$\frac{\partial \overline{u}}{\partial t} - f_0\overline{v} = -\overline{(u'v')}_y \tag{21.11}$$

Note that on the left-hand side of the equation, there is an x-averaged meridional velocity. From our earlier scaling exercise, we recognize that this in an order Rossby number velocity that remains in the equation, because it is multiplied by the relatively large Coriolis parameter and is thus of the same order as the (weak) acceleration of the $O(1)$ geostrophic velocity. The adiabatic equation when x-averaged is

$$\frac{\partial \overline{\rho}}{\partial t} - \overline{w} \frac{\rho_0 N^2}{g} = -\left(\overline{v' \rho'}\right)_y \tag{21.12}$$

A superficial glance at these equations appears to suggest that the mean zonal momentum is actually only altered by the Reynolds stress provided by the wave field, while the change in the mean density is associated with the x-averaged wave flux of perturbation density. If this were the case, it would not be consistent with our earlier view that it is the wave pv flux that is responsible for the change of all quantities in the mean state. How can we resolve that apparent discrepancy?

It is important to note that the eddy fluxes as written drive not just the mean u and ρ but also the x-averaged v and w, i.e., the mean circulation in the y-z-plane. Indeed, in principle it is possible that the wave fluxes might produce a balancing meridional circulation with *no change* in the mean zonal velocity and density. There are problems where that is the case, and we shall shortly see how we can predict that. So, the above formulation is not quite a precise enough picture.

We can attempt to deal with the possibility mentioned above by splitting the mean vertical velocity into a part that may be balanced by the wave flux of density plus a *residual circulation*, which we will indicate with an asterisk, i.e., we write

$$\overline{w} = \frac{\partial}{\partial y}\left(\overline{v' \rho'}\right) \frac{g}{\rho_0 N^2} + w* \tag{21.13}$$

and we define $w*$ as the *residual mean vertical velocity* (residual in the sense that it is the mean vertical velocity after having accounted for what may be the purely wave driven part. In terms of which, the adiabatic equation becomes

$$\frac{\partial \overline{\rho}}{\partial t} - w* \frac{\rho_0 N^2}{g} = 0 \tag{21.14}$$

In this formulation, the change of the mean density field is due entirely to the vertical velocity in the residual meridional circulation. If there had been a non-adiabatic source term for density on the right-hand side of the density equation, it would be the residual vertical velocity that would balance that heating or cooling term in the steady state for the mean. Experience has shown that in the presence of the time varying wave fields, it is the residual velocities that most closely resemble the Lagrangian pathways of the fluid in the meridional plane.

We would also like to define a *residual mean meridional velocity*, and here we have to be a bit careful. When the x-average of the continuity equation is taken, we have, in terms of original variables,

$$\overline{v}_y + \overline{w}_z = 0 \tag{21.15}$$

That is, the mean circulation is nondivergent in the y-z-plane. If we are substitut-ing the residual circulation velocities v^* and w^* for the x-averages of v and w, we would like to make sure that v^* and w^* also satisfy the same divergence-free condition. That suggests defining

$$\bar{v} = -\left(\frac{\overline{gv'\rho'}}{\rho_0 N^2}\right)_z + v^* \tag{21.16}$$

for then it follows that

$$v^*_y + w^*_z = 0 \tag{21.17}$$

If this definition for v^* is used in the x-averaged momentum equation,

$$\frac{\partial \bar{u}}{\partial t} - f_0 v^* = -(\overline{u'v'})_y - \left(\frac{\overline{g\rho'v'}}{\rho_0 N^2}\right)_z = \nabla \cdot \vec{F} \tag{21.18}$$

Thus, now the forcing term due to the wave flux in the x-momentum equation is simply the divergence of the EP vector, or as we have seen, the wave pv flux. This is a promising advance, since we anticipate that the changes in the mean fields are given entirely in terms of the pv flux. Note, however, that the divergence of the EP vector drives not only the time derivative of the mean zonal velocity but also the mean re-sidual meridional velocity. How can we sort out one from the other?

The mean density equation, as derived above is

$$\frac{\partial \bar{\rho}}{\partial t} - w^* \frac{\rho_0 N^2}{g} = 0 \tag{21.19}$$

Let's take advantage of the thermal wind relation as applied to the mean flow, i.e.,

$$f_0 \bar{u}_z = \frac{g\bar{\rho}_y}{\rho_0} \tag{21.20}$$

Take the z-derivative of the mean momentum equation and the y-derivative of the mean adiabatic equation to obtain

$$-f_0^2 v^*_z + w^*_y N^2 = f_0 \frac{\partial}{\partial z} \nabla \cdot \vec{F} \tag{21.21}$$

Since the residual velocities are nondivergent in the y-z-plane, they can be written in terms of a stream function:

$$v^* = -\chi_z, \quad w^* = \chi_y \tag{21.22}$$

which automatically satisfies the continuity equation for the residual velocities. This in turn leads to the elliptic problem for the stream function:

$$\chi_{yy} \frac{N^2}{f_0^2} + \chi_{zz} = \frac{1}{f_0} \frac{\partial}{\partial z} \nabla \cdot \vec{F} \tag{21.23}$$

Therefore, if the EP vector is known, this elliptic problem (with appropriate boundary conditions) can be inverted to find $\chi(y,z)$ and the residual velocities. Note again that it is determined entirely in terms of the divergence of the EP vector. Once the residual circulation is known, the x-momentum equation yields the change in the mean zonal momentum and the mean density. Of course, this inversion is no simpler than the inversion of the original x-averaged pv equation:

$$\frac{\partial \overline{q}}{\partial t} = -\frac{\partial}{\partial y}(\overline{v'q'}) \tag{21.24a}$$

with

$$\overline{q} = \overline{\Psi}_{yy} + \frac{\partial}{\partial z}\left(\frac{f_0^2}{N^2}\frac{\partial \overline{\Psi}}{\partial z}\right) \tag{21.24b}$$

which can be obtained by taking the y-derivative of the x-momentum equation and the z-derivative of the density equation, which is nothing more than a re-derivation for the mean fields of the potential vorticity equation. This more indirect approach has the conceptual advantage of showing in detail how the mean field changes as a consequence of the wave fluxes. It does not change, indeed it emphasizes, the fact that the change comes about only due to fluxes by the waves of potential vorticity. What then, returning to our first question, can we say a priori about the wave flux of potential vorticity?

If the wave amplitude is small so that the waves satisfy linear pv dynamics, we could suppose that we would calculate the wave field from the linear equation:

$$\frac{\partial q'}{\partial t} + U\frac{\partial q'}{\partial x} + v'\frac{\partial \overline{q}}{\partial y} = \mathrm{Diss}(q') \tag{21.25}$$

Here I have added on the right-hand side of the equation a dissipation term, of arbitrary form, for potential vorticity assuming only that it is linear in q' and has zero x-average. We shall shortly see why this might be an interesting addition to the dynamics.

To find the meridional pv flux, we multiply the above perturbation equation by q' and average in x to obtain

$$\overline{v'q'}\frac{\partial \overline{q}}{\partial y} = \overline{q'\mathrm{Diss}(q')} - \frac{\partial \overline{q'^2/2}}{\partial t} \tag{21.26}$$

The potential vorticity flux, when x-averaged, is therefore proportional to the average increase with time of the variance of the wave pv and to the correlation of the pv with its own dissipation. *For steady, inviscid waves, both terms will be zero and the wave pv flux will vanish.* In this case, it follows immediately that *there will be no change in the mean zonal velocity or density fields due to the waves.* The mean residual circulation will be zero. There can be a mean Eulerian circulation

$$\overline{v} = -\left(\frac{\overline{gv'\rho'}}{\rho_0 N^2}\right)_z \quad \text{and} \tag{21.27}$$

$$\overline{w} = \frac{\partial}{\partial y}\left(\overline{v'\rho'}\right)\frac{g}{\rho_0 N^2}$$ (21.28)

so that the wave fluxes of momentum and density yield only a balancing Eulerian v and w but no change in the zonal velocity or its supporting density field. We saw something like this when we looked at the steady internal wave field radiated by the interaction of a current with a rippling topography, and we argued that the only change in the current would occur at the front of the radiating wave field where the time dependence of the wave envelope would be strong. We see here a similar situation for geostrophic flow. This was first noticed, with some expression of amazement, by Charney and Drazin (1961) in their pioneering paper on the propagation of planetary waves from the troposphere into the upper atmosphere. They carefully calculated the wave field and its effects on the mean field and found the effect was zero. Since that time, a good deal of effort has gone into sharpening the theory to describe in detail the role of dissipation and time dependence in describing how the waves *can* affect the mean. A good example of this is found in a very nice paper by Edmond, Hoskins, and McIntyre (1980). The resulting theory is by now rather vast, and further discussion is beyond the scope of this course.

Further efforts to develop the theory for more oceanographically pertinent situations attempt to replace the zonal average (not terribly apt for the ocean) with a time average. The resulting equations are complex, and it is still hard to see clear conceptual progress.

Problems

Problem Set 1

1. As discussed in the text we can consider the generalization of a plane wave to have the form for waves of *slowly varying properties*;

$$\psi = A(\vec{x})e^{i\Theta(x,y,z,t)}$$

and we have defined the wave number vector as the gradient of the phase Θ.

a Show that the condition, for example, that the x-wave number is slowly varying (i.e., that the local definition of a wave number makes sense) is that:

$$\frac{\Theta_{xx}}{\Theta_x^2} \ll 1$$

and carefully interpret this result, i.e., what does the condition mean and why should the condition be imposed? Do the same for the frequency.

b Consider a circular water wave, perhaps formed by a stone thrown in a pond, whose free surface elevation is given by:

$$\eta = A\sqrt{\frac{r_0}{r}}\, e^{i(\kappa r - \omega t)}$$

where r_0 is a constant and r is the circular radius:

$$r = (x^2 + y^2)^{1/2}$$

Assuming the wave is slowly varying, find the x- and y-wave numbers of the wave field at each point in the x-y-plane.

c Using your results in (a), under what circumstances will the assumption in (b) be sensible? This should depend on κ and r.

2. The dispersion relation for Rossby waves which we will derive later, might be approximated as

$$\omega = -\frac{\beta k}{(k^2 + l^2)}, \quad \omega \le 0$$

where k and l are the x- and y-wave number components. β is a parameter. For the planetary problem, it can be shown that β is a measure of the Earth's rotation and sphericity or it also could be related to the slight slope of the bottom of the fluid (as we shall see).

Let's suppose it is the latter, and in our lab we have let

$$\beta = \beta(y) = \beta_0(1-[y/L]^2)$$

where L is large compared to a wavelength of the wave.
a Find the x- and y-components of the group velocity.
b Derive the ray equations for the variation for k, l and ω.
c Show that along the ray path k and ω are constant so that

$$l = \left[\frac{\beta_0(1-y^2/L^2)k}{-\omega} - k^2 \right]^{1/2}$$

d Find the position y_0 where the group velocity in the y-direction vanishes. Note from the ray equation for l that l continues to decrease at that point (i.e., becomes negative). Discuss the implications of that for the trajectory of a wave packet which initially starts near $y = 0$. Sketch the path in the x-y-plane.

3. A particular wave has the form

$$\phi = Ae^{i\theta(x,t)}$$

$$\theta = -gt^2/4x$$

a What is the x-wave number?
b What is the frequency?
c Under what conditions is it sensible to talk about a slowly varying frequency?
d At what speed need you move to see a constant frequency and wave number?
e Moving at that speed, what is the relation between frequency and wave number?
f At what speed do you have to move at to see a constant phase, θ, (i.e., stay on a particular crest)? Is that speed constant with time?

Problem Set 2

1. Acoustic waves in their pure form are small, adiabatic perturbations of a medium of otherwise uniform density and pressure. Assuming that the specific entropy can be written as $s = s(p, \rho)$, show that the governing equations of inviscid motion for disturbances propagating in the x-direction are

a $$\rho_0 \frac{\partial u}{\partial t} = -\frac{\partial p}{\partial x}$$

$$\frac{\partial \rho}{\partial t} = -\rho_0 \frac{\partial u}{\partial x} \quad \text{and}$$

$$\left(\frac{\partial s}{\partial \rho} \right)_0 \frac{\partial \rho}{\partial t} = -\left(\frac{\partial s}{\partial p} \right)_0 \frac{\partial p}{\partial t}$$

where 0 subscripts denote variables in the uniform, unperturbed state.

b Then show that p satisfies

$$\frac{\partial^2 p}{\partial t^2} = c_a^2 \frac{\partial^2 p}{\partial x^2}$$

Identify the sound speed c_a and discuss the nature of the solutions of the equation. Do signals disperse? What significance does this have for communication by speech?

For a perfect gas like air, $p = \rho RT$, and under adiabatic transformation it follows from the standard thermodynamic relations that $(\partial p / \partial \rho)_s = \gamma RT$, $\gamma = c_p / c_v$. What is the sound speed at room temperature?

2. Consider the atmospheric pressure field

$$p_a = P_0 \cos(kx - \sigma t)$$

moving over an infinite body of water of depth D. Find the resulting periodic solution of the water after all initial transients have decayed.

3. Consider a small circular pond of depth D. Suppose the radius of the pond is R. Find the free modes of oscillation for the free surface under gravity. Be sure to carefully state the boundary conditions at the lateral boundary of the basin. Which mode has the lowest frequency? If $D = 3$ meters and R is 10 meters, find that frequency. What is the corresponding frequency in a small water glass (give an estimate)?

(Hint: Find solutions in the form $\phi = F(r) \cosh K(z + D) e^{im\theta} e^{-i\omega t}$, and you may be surprised to discover which m yields the lowest frequency.)

Problem Set 3

1. For a plane gravity wave of the form

$$\eta = \eta_0 \cos(kx - \omega t)$$

we assumed in the text that we could neglect (a) nonlinearity, (b) friction, (c) compressibility, and (d) planetary rotation.

Check these assumptions and discuss, in each case, what *non-dimensional parameter* measures the goodness of the approximation. Make sure you write the condition in terms of quantities given in terms of η_0, k, D and properties of the fluid such as g and v. Be careful to distinguish the conditions when kD is both large and small. You may use sensible values of the wavelength, depth, etc. to get an idea of what limits these parameters set.

2. Consider a rectangular tank of sides L_x, L_y and depth D filled with homogeneous, incompressible fluid. Suppose the fluid in this small basin is forced by a surface pressure of the form

$$p_a = P_0 \cos(kx - \sigma t)$$

Find the linear, forced solution (note: Since the problem is linear, the response must be oscillating at the forcing frequency, but the spatial structure will be modified by the geometry of the basin). Be sure to carefully pose the boundary conditions on the side walls. What Fourier series in x and y is appropriate for the boundary conditions?

When will resonance occur?

What is the solution for small σ?

What do we mean by small σ?

3. In class we derived an energy equation for a layer of fluid supporting gravity waves in the case when the applied atmospheric pressure was zero. Redo the calculation when $p_a \neq 0$.

Problem Set 4

1. Consider the motion of a homogeneous layer of fluid of constant density and of depth D. At $t = 0$, the surface of the fluid is flat but there is a vertical velocity such that

$$w(x, z = 0) = W_0(x)$$

Formulate the initial value problem and find the solution for $\eta(x, t)$ in terms of a Fourier integral and discuss the solution *without* reproducing the details of the derivation of the stationary phase argument.

2. Energy in the internal gravity wave frequency range is generated at $z = z_0$ with an x-wave number k and a z-wave number m.

   a Find the path of such a packet of energy in the x-z-plane (i.e., find dz/dx for the group velocity ray). Estimate the time it would take the packet to reach a depth D if it starts near the surface and if you assume N is independent of z.

   b Discuss how you would do the problem in part (a) if $N^2 = N_0^2 \exp(z/d)$, where d is the thermocline scale (about 1 000 meters) and the vertical wavelength of the gravity wave is much less than d.

3. Consider a stratified fluid with constant N in an infinitely long channel of width L with a rigid lid. Suppose that at $t = 0$,

$$w = w_0(x) \sin \pi z / D$$

$$w_t = 0$$

Find the solution of the initial value problem if w_0 is an even function of x (hint: Note that with the initial condition as given, a solution for all $t > 0$ can be found in the form $w = W(x, t) \sin(\pi z / D)$). *Qualitatively* discuss the solution after you have obtained it.

Problem Set 5

1. Consider a plane, internal gravity wave in a fluid with a constant buoyancy frequency. Calculate both the kinetic and potential energies and discuss whether there is equipartition of the energy (note, it is convenient to choose a coordinate system so that the wave vector lies in the x-z-plane).

2. The vorticity components along the three coordinate axes are

$$\omega_x = \frac{\partial w}{\partial y} - \frac{\partial v}{\partial z}$$

$$\omega_y = \frac{\partial u}{\partial z} - \frac{\partial w}{\partial x}$$

$$\omega_z = \frac{\partial v}{\partial x} - \frac{\partial u}{\partial y}$$

a   Derive, from the linear equations of motion, equations for the rate of change of these vorticity components. In particular, show how the horizontal gradients of density produce vorticity and physically interpret your result.
b   Calculate the vorticity in a plane internal gravity wave when N is constant.

3. Consider the reflection of an internal gravity wave from a sloping surface. Show that the energy flux *normal* to the surface of the incident wave is equal to the energy flux of the reflected wave. We showed in class that the energy densities of the incident and reflected waves were not equal. Is energy conserved?

Problem Set 6

1. Reconsider the normal mode problem for internal waves in the case where $N^2 < 0$, i.e., *when heavy fluid is initially on top of lighter fluid* so that $\partial \rho_0 / \partial z > 0$. Let the fluid be contained in a layer of depth D between two rigid surfaces and let N^2 be constant.
a   What are the frequencies of the normal modes? Are they real? Interpret your result in terms of growth of the disturbance.
b   For what wavelengths will the perturbations grow the fastest?
c   Given the length scale for maximum growth rate, what effect do you think friction or heat conduction would have in determining the wavelength of maximum growth?

2. Calculate the normal modes of internal gravity waves for a stratified fluid with $N^2 > 0$ when the fluid is contained in a box with sides of length L_x and L_y and with depth D. You may assume the upper boundary is a rigid lid. Find the free modes of oscillation and their frequencies (hint: The boundary condition on $x = 0$, say, is $u = 0$. That implies that $\partial p / \partial x = 0$ there for *all* z and so that $(\partial / \partial x)(\partial p / \partial z) = 0$ on $x = 0$. You can use that to write the condition in terms of w).

3. Find an expression for the frequency leading to *critical* angle reflection of internal gravity waves in terms of the local value of ω, N and the slope of the topography. Calculate the condition on frequency for realistic oceanic values. Do you believe the Coriolis effect can be ignored for such frequencies? How would you decide?

Problem Set 7

1. Consider a layer of fluid with a buoyancy frequency N (constant). The fluid is flowing in the positive x-direction with constant velocity U. The base of the fluid is rippled such that

$$h = h_m \cos kx$$

where h is the (small) departure of the bottom of the fluid from a flat surface. *The upper surface of the fluid is level and rigid at a distance D from the bottom.*
a Find the steady solution for the flow (it is nonrotating).
b Discuss whether resonance can occur and interpret your result.
c Calculate the drag on the rippled boundary. Are you surprised (hint: Consider an explanation in terms of the net radiation of energy and the relation between work done and drag)?

2. A channel, *semi*-infinite in the x-direction $(0 \leq x \leq \infty)$ of depth D and width L $(0 \leq y \leq L)$, contains a stratified fluid of constant buoyancy frequency N. The fluid is contained between two level, *rigid* horizontal boundaries. At $x = 0$, a wave maker continuously imparts to the fluid a velocity in the x-direction,

$$u = R_e U_0 \cos (\pi z / D) e^{i\omega t}$$

Find the periodic response of the fluid to the periodic forcing. If ω is less than N, carefully describe how you determine the proper condition on the solution for large x (hint: You may have to apply a radiation condition).

3. Consider a plane wave in an unbounded, stratified fluid with constant buoyancy frequency N and constant Coriolis parameter f. For simplicity, align the x-axis of the system so that the wave vector has no y-component.
a Is the energy equipartitioned between kinetic and potential energy? Is there a particular wavelength for which equipartition occurs?
b In terms of the stream function amplitude, determine the vorticity and the potential vorticity.

Problem Set 8

1. Consider the motion of a rotating, homogeneous layer of water of depth D. Let the layer be infinite in horizontal extent. Suppose that at $t = 0$, the elevation of the free surface above its resting value is given by

$$h = h_0, \quad -a \leq y \leq a$$

and is zero elsewhere. The velocity at $t = 0$ is also zero.

a Find the equation governing the free surface displacement in the steady geostrophic portion of the solution.

b Show that at $y = \pm a$, both η and its first derivative are continuous in the geostrophic solution.

c Find the steady solution for the x-velocity.

d Using the relation between the free surface elevation and the potential vorticity, find the energy in the steady geostrophic state. Discuss, as a function of the ratio of the deformation radius to the length interval $2a$ the percentage of the initial energy radiated away by gravity waves.

2. Show that for the linearized motion of a layer of homogeneous, rotating fluid, that the relation between the free surface height and the velocities can be written:

$$\frac{\partial^2 \vec{u}}{\partial t^2} + f^2 \vec{u} = -g \frac{\partial}{\partial t} \nabla \eta + gf \hat{k} \times \nabla \eta$$

(you may find it useful to write the above in component form).
    What happens for an oscillation for which the frequency exactly matches the Coriolis parameter?

3. Consider a rotating fluid of depth D contained in the region $x \geq 0$, $-\infty \leq y \leq \infty$. Suppose that along a wall at $x = 0$, the velocity in the x-direction is given by

$$u = U_0 e^{i(ly - \omega t)}$$

where it is understood that the real part of the above term is relevant.
    Find the solution for the free surface height in $x > 0$. Distinguish the case when the frequency is greater or less than f.

Problem Set 9

1. Consider the Kelvin wave in a channel of width L. If the free surface elevation has the form

$$\eta = \eta_0 \cos(kx - \omega t) e^{-fy/c_0}$$

a Find the relative vorticity in the wave and its potential vorticity.

b Calculate the kinetic and potential energy in the wave and check for equipartition.

c Discuss the trajectory of fluid elements as the wave passes.

2. A Poincaré wave with x-wave number k_1 (<0) and y wave number ℓ_1 approaches a wall at $x = 0$ from the right.

a What angle does the group velocity make with the x-axis?

b What is the frequency of the wave?

c If the amplitude of the free surface height in the incident wave is A_1, find the complex amplitude of the free surface height of the reflected wave and its x- and y-wave numbers (hint: Be sure to carefully write out the condition $u = 0$ at the wall).

3. a Derive the governing equation for the velocity component, v, in for a layer of rotating fluid of constant depth in the channel as discussed in class and discuss its solutions of the form $v = v(y)e^{i(kx-\omega t)}$.

   b From v, how would you find u and η? For what frequencies does this relation fail? (Hint: To find u in terms of v, take the time derivative of the x-equation of motion and use the continuity equation to eliminate η_t, then use that relation to find η in terms of v)

Problem Set 10

1. Consider a plane Rossby wave with a free surface

$$\eta = \eta_0 \cos(kx + \ell y - \omega t)$$

   a Calculate the kinetic and potential energy in the wave. Check for equipartition. Is there a particular wavelength for which equipartition obtains if it is not true generally?

   b Calculate c_{g_x} as a function of k. Where does it have its largest positive and negative values? Where is it zero? For a wave with a wavelength $\lambda = 50$ km, *estimate* the period of the wave for a fluid of depth 4 km (the precise value will depend on the orientation of the wave vector).

2. a Show that a single plane Rossby wave is an exact solution of the nonlinear quasi-geostrophic potential vorticity equation (qgpve) (hint: First calculate the relative vorticity in the wave and show it is a constant multiple of the stream function).

   b Show that an arbitrary sum of plane Rossby waves will be a solution of the nonlinear qgpve if the *magnitude* of the wave vector of each wave is identical. Note that the frequencies of the waves will differ. Suppose, instead, you have a set of waves of varying wavelengths but whose wave vectors are co-linear?

3. Consider a channel of width L on the beta plane, i.e., $0 \le y \le L$. The bottom is flat. At $x = 0$, a wave maker produces a zonal velocity of the form

$$u = U_0 \cos(\pi y / L)e^{-i\omega t}$$

   a Find the resulting Rossby wave for the region $x > 0$.

   b Do the same for the region $x < 0$.

Problem Set 11

1. Reconsider the development of the quasi-geostrophic equations when

$$\frac{\beta L}{f} = O(1), \quad \frac{h_b}{D} = O(1) \quad \text{and} \quad \varepsilon \ll 1$$

   Derive the governing potential vorticity equation in this limit. Under what circumstances could the Rossby wave frequency satisfy $\omega \ll f$?

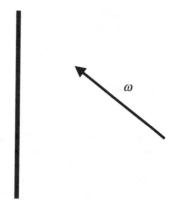

Fig. p.01.
A definition figure for problem 2. The *arrow*
shows the direction of the incident wave energy

2. Consider the reflection of a westward propagating Rossby wave. Its group velocity is directed west-northwestward in a direction that slopes 45° to the northeast from a latitude circle (see Fig. p.01).
   a If the frequency is given, how would you determine the wave number vector of the incident wave?
   b Discuss the reflection of the wave. In what direction is the reflected group velocity? What is the wave vector of the reflected wave?

3. Suppose we model the southern boundary y_s of the Gulf Stream as a rippling surface propagating eastward. We prescribe that boundary as

$$y = y_s + Y_0 \sin k(x - ct), \quad Y_0 \ll y_s$$

   Now consider the oceanic region south of that boundary (i.e., $y \le y_s$).
   Describe the resulting possible wave radiation in the region $y < 0$. Consider both positive and negative values of c.

Problem Set 12

1. Consider the dynamics of a Rossby wave triad as discussed in class. From the properties of the function $P(K_n, K_m)$, show that the *enstrophy* in the triad

$$V = \sum_{j=1}^{3} E_j(K_j^2 + a^2)$$

   is conserved where E_j is the energy in each wave component (note: This implies that

$$\sum_{j=1}^{3} E_j K_j^2$$

   is also conserved).

2. From the quasi-geostrophic equations, show that for a fluid in an infinite region whose motion is limited to the finite part of the x-y-plane, the total enstrophy

$$V = \iint \left(\zeta - a^2 \psi \right)^2 dx dy$$

is conserved.

3. Consider the reflection of a linear Rossby wave from a western boundary oriented in the north/south direction (parallel to the y-axis). Calculate the enstrophy of the incident and reflected waves. Is the emerging enstrophy *flux* equal to the incident flux? If not, what is the mechanism for the nonconservation? Discuss your result and its implications. What is the situation if the reflection occurs at a northern boundary that lies along a latitude circle?

References

Lecture 1

Lighthill J (1978) Waves in fluids. Cambridge University Press, Cambridge, pp 504
Whitham GB (1974) Linear and nonlinear waves. John Wiley & Sons, New York, pp 636

Lecture 2

Bretherton FP (1971) The general linearized theory of wave propagation. In: Reid WH (ed) Mathematical problems in the geophysical sciences, vol 1. American Mathematical Society, pp 61–102
Pedlosky J (1987) Geophysical fluid dynamics. Springer-Verlag, New York, pp 710

Lecture 3

Batchelor GK (1967) An introduction to fluid dynamics. Cambridge University Press, London, pp 615 (especially Chapter 1)
Kundu PK (1990) Fluid mechanics. Academic Press, San Diego, pp 638 (especially Chapter 5)
Lamb H (1945) Hydrodynamics, 6th edition. Dover Publications, New York, pp 738
Stoker JJ (1957) Water waves. Interscience, New York, pp 567

Lecture 4

Kundu PK (1990) Fluid mechanics. Academic Press, San Diego, pp 638 (especially Chapter 5)
Stoker JJ (1957) Water waves. Interscience, New York, pp 567

Lecture 5

Jeffreys J, Jeffreys BS (1962) Methods of mathematical physics. Cambridge University Press, Cambridge, pp 716 (especially Chapters 14 and 17)
Lighthill J (1978) Waves in fluids. Cambridge University Press, Cambridge, pp 504 (especially Chapter 3, Section 3.7)
Morse PM, Feshbach H (1953) Methods of theoretical physics, vol 1. McGraw-Hill, New York, pp 997 (especially Section 4.8)
Stoker JJ (1957) Water waves. Interscience, New York. pp 567

Lecture 6

Rossby CG (1945) On the propagation of frequencies and energy in certain types of oceanic and atmospheric waves. J Meteorol 2:187–204
Stoker JJ (1957) Water waves. Interscience, New York, pp 567
Whitham GB (1974) Linear and nonlinear waves. John Wiley & Sons, New York, pp 636 (especially Section 11.4)

Lecture 7

Gill AE (1982) Atmosphere-ocean dynamics. Academic Press, Harcourt Brace & Co., San Diego, pp 662 (especially Chapter 6)
Lighthill J (1978) Waves in fluids. Cambridge University Press, Cambridge, pp 504 (especially Chapter 4)
Munk W (1981) Internal waves and small scale processes. In: Wunsch C, Warren BA (eds) Evolution of physical oceanography. MIT Press, pp 264–291

Lecture 8

Lighthill J (1978) Waves in fluids. Cambridge University Press, Cambridge, pp 504 (especially Chapter 4)
Mowbray DE, Rarity BSH (1967) A theoretical and experimental investigation of the phase configuration of internal waves of small amplitude in a density stratified liquid. J Fluid Mech 28 1–16

Lecture 9

Gill AE (1982) Atmosphere-ocean dynamics. Academic Press, Harcourt Brace & Co., San Diego, pp 662 (especially Chapter 6)
Levitus S (1982) Climatological atlas of the World Ocean. U.S. Dept. of Commerce, US. Government Printing Office, pp 173

Lecture 10

Eliassen A, Palm E (1960) On the transfer of energy in stationary mountain waves. Geofys Publ 22:1–23
Lighthill J (1978) Waves in fluids. Cambridge University Press, Cambridge, pp 504

Lecture 11

Cairns JL, Williams GO (1976) Internal wave observations from a midwater float, 2. J Geophys Res 81:1943–1950
Garrett C, Munk W (1979) Internal waves in the ocean. Annu Rev Fluid Mech 11:339–369

Lecture 12

Cahn A (1945) An investigation of the free oscillations of a simple current system. J Meteorol 2:113–119
Gill AE (1982) Atmosphere-ocean dynamics. Academic Press, Harcourt Brace & Co., San Diego, pp 662 (especially Chapter 7)
Rossby CG (1938) On the mutual adjustment of pressure and velocity distributions in certain simple current systems, II. J Mar Res 1:239–263

Lecture 13

Cahn A (1945) An investigation of the free oscillations of a simple current system. J Meteorol 2:113–119
Gill AE (1982) Atmosphere-ocean dynamics. Academic Press, Harcourt Brace & Co., San Diego, pp 662
Pedlosky J (1987) Geophysical fluid dynamics. Springer-Verlag, New York, pp 710

Lectures 14, 15

Pedlosky J (1987) Geophysical fluid dynamics. Springer-Verlag, New York, pp 710 (especially Chapter 6)

Lecture 16

Greenspan HP (1968) The theory of rotating fluids. Cambridge University Press, London, pp 327
Longuet-Higgins MS (1964) On group velocity and energy flux in planetary wave motions. Deep-Sea Res 11:35–42
Pedlosky J (1987) Geophysical fluid dynamics. Springer-Verlag, New York, pp 710 (especially Chapters 3 and 4)

Lecture 17

The derivation of the Laplace tidal equations in oceanic and atmospheric contexts can found in many references, among which:

Andrews DG, Holton JR, Leovy CB (1987) Middle atmosphere dynamics. Academic Press, New York, pp 489 (especially Chapter 4)
Moore DS, Philander SGH (1977) Modeling the tropical oceanic circulation. In: Goldberg ED, McCave IN, O'Brien JJ, Steele JH (eds) The Sea, vol 6. Wiley Interscience, pp 319–361
Pedlosky J (1987) Geophysical fluid dynamics. Springer-Verlag, New York, pp 710 (especially Chapter 6)

Lecture 18

Andrews DG, Holton JR, Leovy CB (1987) Middle atmosphere dynamics. Academic Press, New York, pp 489 (especially Chapter 8)
Eriksen CC, Blumenthal MB, Hayes SP, Ripa P (1983) Wind generated Kelvin waves observed across the Pacific. J Phys Oceanogr 13:1622–1640
Moore DS, Philander SGH (1977) Modeling the tropical oceanic circulation. In: Goldberg ED, McCave IN, O'Brien JJ, Steele JH (eds) The Sea, vol 6. Wiley Interscience, pp 319–361
Philander SG (1990) El Niño, La Niña, and the southern oscillation. Academic Press, San Diego, pp 289
Schiff LI (1955) Quantum mechanics. McGraw-Hill, New York, pp 417 (especially Chapter 4)

Lecture 19

Charney JG (1947) The dynamics of long waves in the westerlies. J Meteorol 4:135–162
Eady ET (1949) Long waves and cyclone waves. Tellus 1:33–52
Pedlosky J (1987) Geophysical fluid dynamics. Springer-Verlag, New York, pp 710 (especially Chapter 6)

Lecture 20

Gill AE (1982) Atmosphere-ocean dynamics. Academic Press. Harcourt Brace & Co., San Diego, pp 662
Heifetz E, Alpert P, Da Silva A (1998) On the parcel method and the baroclinic wedge of instability. J Atmos Sci 55:788–795
Pedlosky J (1987) Geophysical fluid dynamics. Springer-Verlag, New York, pp 710 (especially Chapter 7)

Lecture 21

Charney JG, Drazin PG (1961) Propagation of planetary scale disturbances from the lower to the upper atmosphere. J Geophys Res 66:83–109
Edmond HJ, Hoskins BJ, Mcintyre ME (1980) Eliassen-Palm cross sections for the troposphere. J Atmos Sci 37:2600–2616

Index